CLAUDIA WEISBURD has farmed 40 acres for many years, raising pigs, dairy and beef cattle, sheep, poultry, goats, and horses. A graduate of Cornell University College of Agriculture and Life Sciences, she is presently farming an agricultural cooperative in Ithaca, New York.

RAISING YOUR OWN LIVESTOCK

CLAUDIA WEISBURD

A SPECTRUM BOOK

PRENTICE-HALL, INC., Englewood Cliffs, New Jersey 07632

Library of Congress Cataloging in Publication Data

WEISBURD, CLAUDIA.
  Raising your own livestock.

  (A Spectrum Book)
  Bibliography: p.
  Includes index.
  1. Livestock. I. Title.
SF65.2.W44-1980   636   80-402
ISBN 0-13-752758-6
ISBN 0-13-752741-1 pbk.

To Jerry, who learned to like animals for my sake, and to Kate, who may learn to love them on her own, once she knows what they are.

The author gratefully acknowledges the contribution through classes, personal conferences, and extension publications, of the New York State College of Agriculture and Life Sciences, Cornell University.

Line drawings by Jerry Weisburd

Interior design and production by
Nancy Earle
THE BOOK DEPARTMENT
52 Roland Street
Charlestown, Massachusetts 02129

Editorial/production supervision by Norma Miller Karlin

Manufacturing buyer: Cathie Lenard

Printed in the United States of America

PRENTICE-HALL INTERNATIONAL, INC., *London*
PRENTICE-HALL OF AUSTRALIA PTY. LIMITED, *Sydney*
PRENTICE-HALL OF CANADA, LTD., *Toronto*
PRENTICE-HALL OF INDIA PRIVATE LIMITED, *New Delhi*
PRENTICE-HALL OF JAPAN, INC., *Tokyo*
PRENTICE-HALL OF SOUTHEAST ASIA PTE. LTD., *Singapore*
WHITEHALL BOOKS LIMITED, *Wellington, New Zealand*

CONTENTS

I.
SOME BASIC ANIMAL SCIENCE EVERY LIVESTOCK OWNER SHOULD KNOW   1

    INTRODUCTION   2
1. NUTRITION   4
2. ANIMAL BREEDING   20

II.
DAIRY ANIMALS   35

3. DAIRY COWS   41
4. DAIRY GOATS   91

III.
RAISING MEAT ANIMALS   111

5. MEATS   112
6. BEEF CATTLE   126
7. SWINE   157
8. SHEEP   192

IV.
POULTRY 221

9. CHICKENS 222

V.
HORSES 251

10. HORSES 252

VI.
OTHER MANAGEMENT 281

11. THE BARN 282
12. FIRST AID 290

APPENDICES

  I. USING FEEDING STANDARDS 297
 II. HAY, PASTURE, AND SILAGE 302
III. LIVESTICK DATA SUMMARY 311

INDEX 312

I.
SOME BASIC ANIMAL SCIENCE EVERY LIVESTOCK OWNER SHOULD KNOW

1. NUTRITION
2. ANIMAL BREEDING

Introduction

The great American way of life: riding lawn mowers, drive-ins, drive-throughs, electric knives, and exercisers to keep fit; big government, big business, big buildings, big cars. Is it any wonder that so many people are trying to simplify and decentralize? Or that they are becoming concerned with the scale of their lives and their loss of contact with the basic aspects of survival? Food, shelter, clothing: how many people are *directly* involved with providing these necessities for themselves? Perhaps for this reason, because things have gotten too big and too removed, people are trying to do more for themselves, on a small and human scale. Backyard gardens, making clothes, baking bread, all are ways that people are scaling down and getting in touch with providing for themselves.

Much of this "back to basics" drive has centered on food, from the interest in health food to the revival of small farmers' markets to more and more people growing their own. Interest in vegetable gardening has burgeoned over the past several years as supermarket prices rose, quality declined, and people grew dissatisfied with their lack of control over what they could get to eat. There is no reason to stop with vegetables, however; livestock products can also be produced on a small scale, and many people are now trying to do just that, keeping some backyard livestock to provide meat, dairy products, and eggs.

Small-scale stockkeeping can be enormously rewarding: the reward of halved grocery bills, the reward of the superior quality of the fresh products, and most of all, the reward of being involved with the animals themselves. There are few sounds more peaceful than that of contented animals munching their evening meal, and few feelings as good as knowing you are giving something to the animals you're getting so much from. A relationship develops: caring, warmth, respect for the animals who are providing for you. And also—it's fun. If you give them (and yourself) half a chance, not only will your livestock provide you with food but also with hours of entertainment. Pigs, chickens, goats, sheep, cows—they all do ridiculously funny things, if you just take the time to slow down and watch them.

But don't fool yourself. Keeping livestock requires a whole lot more responsibility than a vegetable garden; animals are, after all, living beings, dependent on you. They need to be housed properly and fed regularly. They cannot be left to fend for themselves over weekends and holidays. They cannot be sloppily and carelessly tended. The person taking on livestock is taking on work—often hard, tedious, and dirty work—and the responsibility to work consistently and correctly. And that's what this book is about. Naturally, the best way to learn is to do, but it's better for everyone if you can start off on the right foot. It's always nice, too, to have some help along the way, and even nicer to be able to get some idea of what you might be getting into before taking a blind plunge you might regret later.

Yes, be forewarned that the work and the responsibility are not to be taken lightly. But also realize that you cannot overestimate the pleasure involved in small-scale stockkeeping. Then, too, there's the satisfaction of sinking into bed after a hard day, knowing that you are an integral, active part of the food chains and processes that make up your life.

1
NUTRITION

"Everybody's heard how rugged beef cattle are, so people think they can stick them outside and let them live on scenery and the north wind," a cooperative extension agent once complained. An owner of a small dairy was lamenting her hay situation: "I just can't find any good hay. These cows had an 11,000-pound average in their last herd; now they're down to 8,000 pounds. But what do you expect? Look at what they're eating!" Some owners of a few backyard pigs said, "I don't know how other people do it, but we just can't seem to get our pigs to grow to slaughter weight by six months." What is the root of all these problems? It's nutrition, and good nutrition is one of the major keys to successful animal husbandry. Almost anyone can tell you what makes a balanced meal for a person, but when it comes to feeding livestock, many small-scale stock owners are adrift in a sea of misinformation, old wives' tales, and just plain ignorance. Some understanding of the basic principles of nutrition is vital for intelligent management and will help immeasurably in dealing with specific animals in specific situations.

Feeding for Production

The first question to ask concerning feeding is "What does it take to simply maintain the animal?" A maintenance ration is fundamental; it just meets the requirements of an animal who is doing no work and is producing nothing, and it is considered to be proportional to the animal's body weight.

The next question is "What is the animal doing, or producing?" Rarely, I would hope, is the answer, "Nothing." "Nothing" would indicate a maintenance ration, but the production of *anything* will require some feeding in addition to the basic nutritive level of maintenance. "Production" can be thought of as growth (the animal is producing itself), reproduction, fattening, lactation, work, egg production, or sometimes, combinations of these (i.e., growth and fattening in pigs, lactation and reproduction in dairy animals, or growth and reproduction in pregnant heifers).

In general terms, the requirements for different types of production are as follows:

Growth requires a high-energy, high-protein diet and careful attention to vitamins and minerals. Feeding during the growing period is especially important economically for meat producers because young animals gain more quickly and efficiently than older ones. The higher efficiency of early gains is related to several factors:

1. Early weight gains are higher in water than later weight gains, and water is cheap.
2. Early weight gains contain less fat. It takes 2.25 times as much feed energy to put on a pound of fat than a pound of lean, and energy is not so cheap.
3. Young animals consume a larger amount of feed per unit of body weight than do older animals, which means that after the maintenance requirements have been met, there is still a large proportion of nutrients left over for production.

It is easy to see why it pays for pig producers, for example, to get the weight on the pigs and the pigs off to market as soon as possible.

Reproduction, although not as demanding as growth, does require a higher plane of nutrition than the maintenance level. Feeding must be adequate for the provision of nutrients for the developing fetus and membranes, as well as for the later production of milk. The nutritive requirements of pregnancy increase most during the last third of the gestation period, which is the time of greatest fetal weight gains. In early pregnancy, energy levels can be kept at as little as 10 percent above maintenance, but in later pregnancy, requirements jump to an additional 30 percent. Protein, too, follows suit: protein requirements in early pregnancy are only slightly above maintenance levels, whereas during the last third of the gestation period, they are 50 percent above maintenance. All this information pertains to the female; what of the other component of the reproductive process? Males used for occasional service can be merely maintained, but if they are used for regular service, they will need an increase in energy levels of about 20 percent.

Fattening is often considered to be the feeding of a mature animal so it lays down fat. The cost of this weight gain is relatively high, both because of the higher energy required for laying down fat, as explained earlier, and because the basic maintenance requirement increases as the animal gets heavier. The primary requirement for fattening is energy; protein levels needed are only slightly above maintenance. Of the nutrients, carbohydrates, fat, or protein can all be used by the body to produce fat, but for practical purposes, carbohydrates are the primary ingredient of feeds used for fattening. Interestingly enough, though, some feeding of fats can change the fat in the animal. For example, if pigs are fed large quantities of peanuts, which are high in the unsaturated, soft fat, oleic acid, soft body fat in the pig will result—not a desired effect. On the other hand, unsaturated (soft) fats are saturated (hardened) in the rumen of a cow, so feeding soft fats will not generally affect milk fat. If, however, really large amounts of unsaturated fat are fed to a dairy cow, it is possible to end up with "soft butter."

Frequently, animals are being fattened as they are growing (finishing pigs or lambs), and so a high level of nutrition is needed to meet the energy and protein needs of growth as well as the added energy needs of fattening.

Lactation demands are quite high and increase with increasing levels of milk production. The components of milk are largely synthesized in the udder from precursors in the blood, although vitamins and minerals filter through directly. It is the products of digestion (amino acids, sugars, fatty acids, *etc.*) that act as these precursors; therefore, they must be supplied in adequate amounts in the diet. The effect of the diet on actual milk composition, however, as opposed to its effect on milk quantity, is slight: the percent protein in the milk cannot be radically changed by varying protein levels in the ration; the percent lactose (milk sugar) is constant; calcium and phosphorous levels cannot be changed; some trace elements, such as iron and cobalt, can be increased in milk by increasing feed levels; and the Vitamin A and D content of milk is affected by the vitamin content of the feed.

Work is primarily an energy-requiring function, which of course varies with the amount of work. Although energy levels must increase with increased work, protein requirements for an animal doing heavy work are only slightly above maintenance levels, as are vitamin and mineral requirements.

Egg production requires high energy as well as high protein levels. Another striking need in egg production is for additional minerals, particularly calcium and phosphorous. When you think that the dry matter of milk contains 1 percent calcium (and milk is a good calcium source), whereas the dry matter of an egg contains 15 percent calcium, it is easy to see why the hen has such a high requirement for this mineral. Vitamin requirements for layers are not well established, but it has been seen that the vitamin content of the egg can be increased by feeding high proportions of vitamins in the rations.

An idea of what the feed requirements are for different types of production brings up the question of how to meet them. The first consideration is the different classes of nutrients and their functions, then how the nutrients are handled by different digestive systems, how nutrient levels in feeds are measured, and finally, the types of feeds that will supply the necessary nutrients.

Classes of Nutrients

Carbohydrates, basically considered an energy source, comprise three-fourths of the dry matter of plants: the fibrous cell walls, tubers, and seeds all contain large amounts of carbohydrates; but carbohydrates form only 1 percent of the animal body. Carbohydrates are made up of carbon, hydrogen, and oxygen, with hydrogen and oxygen in a 2:1 ratio. During photosynthesis, plants use carbon dioxide from the air, water from the soil, and energy from the sun to form glucose, a carbohydrate and the basic unit of many compounds. Carbohydrates can be classified as sugars or polysaccharides. The sugars are sweet, are soluble in water, and form crystals readily. The sugars include (among others) the simple sugars glucose, galactose, and fructose, as well as the compound sugars such as sucrose (glucose plus fructose) and lactose (glucose plus galactose).

Polysaccharides are molecules of simple sugars joined in chains. They're almost tasteless, rarely crystallize, and are variably soluble in water. One of the largest groups of polysaccharides consists of glucose chains—starch. Starch is the chief storage component of plants and comprises 70 percent of most cereal grains. It is easily digested and so has a high feeding value. Similar to starch is glycogen—the animal version of starch. Glycogen, also formed of glucose chains (but connected differently), is stored in the muscle and liver and is used as an energy reserve. Cellulose is another complex glucose compound that makes up the cell walls of plants. Cellulose is much more difficult to digest than starch and so has a lower feeding value, particularly for simple-stomached animals. Hemicellulose is similar to cellulose but can be broken down more completely during digestion, giving it a higher feeding value than cellulose.

Proteins contain carbon, hydrogen, oxygen, and nitrogen, and are large complex molecules composed of chains of amino acids. Amino acids are compounds that contain nitrogen and an organic acid; twenty-five amino acids have been isolated. Although each protein has a specific composition, with twenty-five amino acid building blocks to arrange and rearrange, there are innumerable variations, allowing for the fact that proteins form most of the dry matter of tissues and make up enzymes, hormones, antibodies, and hemoglobin.

Although all twenty-five of the amino acids are physiologically essential to the body, only some are essential in the diet; essential amino acids are those that must be eaten, whereas the remaining ones can be synthesized by the body. The number of amino acids essential in the diet varies among species; for most domestic animals and for humans, there are approximately ten. The presence or absence of these essential amino acids in a feed, in the proper amounts and proportions to form the necessary body proteins, determines the quality of the protein fed. A high-quality protein is one that contains the proper amounts of all the essential amino acids. When feeding livestock (or ourselves), attention must be given, therefore, not only to the *quantity* of protein in the diet but also to its *quality*; that is, there must be a certain amount of protein in the diet, and it must contain all the essential amino acids. Most plant products are deficient in one or two amino acids; for this reason, different plants must be combined to achieve the quality protein required. Animal products, on the other hand, usually contain high-quality proteins with the necessary amino acids in the right proportions.

Fats, or lipids, also contain carbon, oxygen, and hydrogen, but with a much lower proportion of oxygen than the carbohydrates. They are soluble in ether or alcohol but not in water. The breakdown of fat releases a great deal of energy—true fats supply 2.25 times as much energy as carbohydrates—and so they are regarded as primarily an energy source in livestock feeding.

Minerals are needed by the body in varying amounts and are important in a number of body functions, including maintenance of pH, osmotic pressure, and enzymatic activity. Generally, minerals are divided into the macroelements and the microelements (trace minerals), depending on the amount needed by the animal. Macroelements (needed in relatively large quantities) are calcium, sodium, potassium, phosphorous, chlorine, sulfur, and magnesium. Trace minerals include iodine, zinc, manganese, cobalt, copper, iron, molybdenum, selenium, and fluorine. Although many of the macroelements and the trace minerals are present naturally in both forages and concentrates, they are usually added to commercial feeds, fed as supplements, or made available in trace mineralized salt blocks (which no animal should be without) to ensure adequate amounts in the diet.

Vitamins, like minerals, and often with minerals, are critical in body functioning, although they are needed in minute amounts. Usually they are considered to function as expediters in normal physiological activities—as catalysts, in metabolic regulation, and

Salt must always be available to your livestock.

in biosynthesis of tissues. For example, Vitamin E is involved in oxidation-reduction reactions; Vitamin D influences phosphorous and calcium metabolism; Vitamin A affects the photoreceptors of the eye; and Vitamin K functions in blood clotting. Vitamins A, D, E, and K are fat soluble; the B vitamins and Vitamin C are water soluble. Vitamins A (found in green plants as carotene), D (from the sun), and E (found in the germ of cereal grains) are the most often considered in livestock feeding. (Table 1-1 gives mineral and vitamin sources and functions.)

Water, the largest component of the animal body, functions as a solvent and acts in transport and chemical reactions. It goes without saying that all animals should have access to ample fresh, clean water at all times.

Digestive Systems

When dealing with farm animals, two distinctly different types of digestive systems are involved: the ruminant and the nonruminant, or simple-stomached animal. Simple-stomached animals (pigs, dogs, people) are fed entirely differently from the ruminants (cows, sheep, goats) because of the difference in their digestive systems.

The nonruminant digestive tract consists of the mouth, esophagus, stomach, small intestine (and associated pancreas and gall bladder), a small cecum, large intestine (colon), and rectum. The

Table 1-1. Minerals and Vitamins in Reference to Livestock Feeding

Nutrient	Functions	Deficiency Symptoms	Sources	Practical Application
NaCl (Salt)	maintains osmotic pressure in body fluids; Na and K maintain rhythmic beat of heart; Cl forms HCl in stomach for digestion	depraved appetite; reduced appetite; drop in body weight; drop in milk production	animal products	1% NaCl in concentrate mix per ton
Calcium and Phosphorous	formation of bones and teeth; Ca required for blood clotting; P involved in metabolism, energy transformation	rickets; osteomalacia in older animals; depraved appetite; posterior paralysis in swine	Ca: legumes, limestone, milk, bonemeal P: milk products, cereal grains	feed legume forages, cereal grains; maintain calcium to phosphorous ratio at 1:1 or 2:1
Iodine	component of thyroxine; regulates metabolism	depression of thyroxine production results in listlessness, weight gain; goiter; birth of dead or hairless young	potassium iodide	use iodized salt
Iron and Copper	Iron: constituent of hemoglobin, myoglobin, and many enzymes Copper: required before iron can be incorporated into hemoglobin; constituent of many enzyme systems	Iron: anemia Copper: "steely wool" in sheep; degeneration of central nervous system	Iron: animal products Copper: usually enough in feeds	Milk deficient in iron; use injectable iron when necessary

Cobalt	constituent of vitamin B_{12}	decided lack of appetite; anemia	supplementary cobalt salts if necessary	required by ruminants only
Vitamin A (retinol)	component of visual purple; general metabolism of epithelial cells	night blindness; xeropthalmia (dry eyes); degeneration of testes; degeneration of developing embryos; degeneration of epithelial tissues in general	carotene, yellow corn, green hay, milk fat, fish liver oils	animal manufactures vitamin A from carotene and can store; green plants provided ample; deficiency rare
Vitamin D (sunshine vitamin)	essential in absorption of calcium; involved in formation of calcium and phosphorous into bones and teeth	rickets	ultraviolet light (sun), sun-cured hay, fish liver oils, irradiated yeast	allow sunlight; animals born in winter need supplementary Vitamin D
Vitamin E (tocopherol)	antioxidant	sterility in poultry and lab animals; degeneration of testes; white muscle disease; stiff lamb disease	wheat germ oil, germ of cereal grains	can use synthetic tocopherol when necessary
Vitamin K	necessary for blood clotting	can cause fatal hemorrhages	abundant in green leaves	not a problem

primary agents of digestion are the enzymes secreted along the way, which break down the carbohydrates, proteins, and fats into smaller units (sugars, amino acids, and simple fats or fatty acids) suitable for absorption. The smaller units, along with minerals and vitamins, are absorbed primarily from the small intestine. The undigested residue of the feed moves down the tract to the large intestine, water is absorbed, and the rest is excreted as feces. Nitrogenous wastes from protein breakdown are excreted in the urine.

The ruminant digestive system also starts with the mouth and esophagus, but next it takes a different turn. Feed then moves to the rumen, reticulum, omasum, abomasum, small intestine, cecum, colon, and rectum. The major agent of digestion in the ruminant is not enzymatic action but rather bacterial fermentation, which begins in the rumen. A whole host of microflora inhabits the rumen, and it is these microorganisms, not enzymes, that act on the ingested feed. In the rumen and reticulum, all carbohydrates are digested, including cellulose and hemicellulose (the bacteria of the rumen can digest cellulose, although simple-stomached animals cannot). Also, in the rumen and reticulum much of the protein is broken down, as are fats. Vitamin K, all the B vitamins, and all the essential amino acids are synthesized in the rumen. Protein synthesis occurs by bacterial action, which uses the nitrogen portion of the protein ingested. The microorganisms use the nitrogen (it doesn't matter where it came from; to a certain extent it can even be from nonprotein sources such as urea) to build their own bodies. The bacteria, which are very high-quality protein themselves, are digested by the ruminant in the small intestine.

In feeding a ruminant, you are in essence feeding the bacteria of the rumen, and they in turn supply the nutrients for the animal. The implications of this system are great. First, you don't have to feed B vitamins. Second, you can feed roughages in quantities nonruminants could not consume or utilize. Most important, the quality of the protein fed to a ruminant is not an issue. Since the bacteria are high-quality protein themselves, only *quantity* of protein, not *quality,* is important when feeding a ruminant. It is worthwhile to note here what a wonderful system this really is. A ruminant can consume material indigestible by humans (such as grass), as well as low-quality proteins, and turn them into highly digestible, high-quality protein food for humans: milk and meat.

Proximate Analysis

How does all this information apply when it comes time to decide what to feed? Well, somehow there has to be a measure of these nutrients in the feed. The most common measure is the proximate analysis, which is found on the tags of most commercial feeds.

Table 1-2. Summary of Major Characteristics of Ruminant and Nonruminant Digestion Processes

	FEEDS						
	CARBOHYDRATES		PROTEIN		FATS	MINERALS	Undigestible Residue
	Simple-stomached	Ruminant	Simple-stomached	Ruminant			
Product of digestion	simple sugars (glucose)	simple fatty acids	amino acids	ammonia and carbon fragments	monoglycerides and free fatty acids	various salts	
Location	stomach small intestine	rumen	stomach, small intestine	rumen, 20-40% in small intestine	small intestine		
Absorbed as	simple sugars	fatty acids	amino acids	bacterial protein	fatty acids and monoglycerides	salts	
Used for	energy (excess stored as glycogen or body fat)	energy, synthesizing glucose, body fat	synthesizing body protein, energy, body fat (very low priority)	synthesizing body protein, energy, body fat (very low priority)	body fat, energy	various	
Excreted	carbon dioxide, water	carbon dioxide, methane, water	nitrogenous compounds (urea, ammonia)			chloride, sulfates, phosphates, etc.	feces

Proximate analysis, rather than measuring each individual nutrient, measures groups of compounds that function similarly nutritionally. Groups measured are water, crude protein, crude fat (or ether extract), ash, and crude fiber.

Water is measured simply by weighing the sample, vaporizing the water, then reweighing the sample. The difference is water; the remainder is "dry matter." Feeds differ greatly in their water contents, ranging from 15 to 20 percent in hays, to 30 to 40 percent in silages, and the water content affects the feeding value. For this reason, feed requirements and comparisons of feeds are frequently made on a dry matter basis.

Crude protein is calculated as the total amount of nitrogen (which includes nonprotein nitrogen) multiplied by 6.25 (on the basis that most protein contains 16 percent nitrogen), giving the total *quantity* of protein in the feed, although there is no indication of protein *quality*. The crude protein measure, therefore, is a good indicator of the protein value for ruminants but is less accurate for nonruminants.

Crude fat is determined by extracting the feed sample with ether, which dissolves out all ether-soluble materials, including the true fats. The ether is then boiled off, and the residue is labeled crude fat or ether extract. However, it is not only the fats that will dissolve in ether; the residue will include such things as chlorophyll and waxes. Crude fat, therefore, is not a very accurate measure of the true fat content of a feed; but since most livestock feeds do not contain significant amounts of fat, the nutritional error is not serious.

Ash content is determined by burning the sample at a temperature high enough to oxidize the organic matter. The residue is ash, or the mineral elements. Most livestock feeds are 2 to 5 percent ash.

Last, and least accurate, is the carbohydrate measure. Carbohydrates include the highly digestible sugars and starches as well as the less digestible cellulose and hemicellulose. The analysis attempts to separate these two groups by alternately boiling the sample in weak acid and weak alkalide solutions, trying to simulate the processes of the digestive tract. Theoretically, this technique will separate the fraction of the least digestibility—this residue is called crude fiber. Crude fiber includes most of the cellulose and some hemicellulose. A high crude fiber content indicates a low digestibility. Cereal grains are 2 to 10 percent crude fiber; hay, on the other hand, is 40 to 50 percent.

The nitrogen-free extract (NFE) is meant to reflect the more digestible carbohydrate fraction. It is determined by difference: add all the other values (water, crude protein, crude fat, ash, and crude fiber), subtract from one hundred, and the answer is the NFE. Although this system actually works reasonably well for concentrates because of their high starch content, it is a very poor indicator of the digestible carbohydrate portion of roughages.

Although a more accurate system has been developed to separate the digestible from the nondigestible fiber (the "detergent fiber" method), crude fiber and NFE are still the most commonly used carbohydrate measures. The detergent fiber method gives a more accurate picture of the digestibility of the carbohydrate fraction by dividing the carbohydrate portion of the plant into cell contents, which are almost completely digestible, and cell wall contents, which are only partly digestible.

Total Digestible Nutrients (TDN)

Although proximate analysis measures the amount of dry matter, crude protein, crude fat, ash, crude fiber, and NFE in a feed, other evaluations must be made to determine its actual feeding value. Not all the nutrients in a feed are digested: some nutrient losses occur through the production of feces, urine, gas, and heat. The easiest loss to measure and the one most commonly accounted for is the nutrient loss in feces. Nutrients in the feed are measured; then, after the animal has been fed, the nutrients in the feces are measured. The difference between "nutrients in" and "nutrients out" is considered to be the amount of each nutrient that is actually digested and is, therefore, available for use by the animal. The average percent of each nutrient digested is called the digestion coefficient (amount absorbed divided by the amount in the feed); digestion coefficients for most feeds have been determined. Adding the *digestible* crude protein (total crude protein times the digestion coefficient), digestible crude fiber, digestible NFE, and the digestible crude fat of a feed gives the TDN value, or total digestible nutrient value, of that feed. The TDN value estimates the approximate energy value of the feed, and although not the most accurate measure, is widely used. TDN values vary among feedstuffs—anywhere from 30 to 130 percent (values over 100 percent are the results of high fat contents). Cereal grains are high in TDN (corn is 80 percent), whereas roughages are generally low (hay is 50 percent). An animal at a given level of production has a certain TDN requirement, which is met by combining different types of feeds.

Types of Feeds

Livestock feeds are generally classified into two major groups: roughages and concentrates. Roughages, usually high in crude fiber and low in digestibility (low TDN), include hay, pasture, silage, straw, stover, etc. Concentrates are low in fiber and have a relatively high digestibility (high TDN). Typically, they are high-energy feeds and include grains and grain by-products, seeds, meals, and animal products.

Roughages

Although good pasture is an ideal roughage, through necessity, convenience, or economics, some form of stored roughage is often fed through some part of the year. The quality of stored roughages is affected by several factors: (1) the stage of maturity when the forage was harvested, (2) the species of the plant, and (3) the method of preservation.

In general, it can be said that as a plant matures, the percent of protein decreases, the percent of digestible protein decreases, crude fiber increases, digestible crude fiber decreases, and total TDN decreases. Table 1-3 illustrates this point for the Northeast. From this table, very early harvest looks appealing, but very early cuttings reduce the total yield per acre; a compromise must be worked out between highest yield with the most nutritive value—which in the Northeast is June 15.

Legumes and grasses are the usual categories of plant species considered. Grasses are monocotyledonous, with (usually) simple, parallel veined leaves. Timothy, brome, cereal grains, and many more are members of the grass family. Legumes are dicotyledonous, with compound leaves. They do not depend on the soil for their nitrogen supply; nodules on their roots contain bacteria, which can utilize nitrogen from the atmosphere and make it available to the plant. Alfalfa, clovers, soybeans, peas, etc., are legumes.

Grasses and legumes each have their advantages and disad-

Table 1-3. Relationship Between Stage of Maturity of Plant and Nutritive Content in the Northeast

| | | PERCENT | | | | |
STAGE OF MATURITY	DATE	C.P.	C.F.	D.P.	D.C.F.	TDN
Vegetative	May 28	22.0	27.9	80.0	74.0	72.0
Early head	June 14	13.7	33.7	71.0	68.0	65.0
Early bloom	July 1	9.1	35.0	55.0	48.0	54.0
Late bloom	July 18	6.8	37.2	36.0	48.0	54.0

C.P.   Crude Protein
C.F.   Crude Fat
D.P.   Digestible Protein
D.C.F.   Digestible Crude Fat
T.D.N.   Total Digestible Nutrients

vantages. Because they are higher in protein, much higher in calcium, and are somewhat more palatable than the grasses, legumes have a higher feeding value. In addition, under good growing conditions, legumes will outyield grasses, at the same time helping to maintain soil fertility because of their nitrogen-fixing ability. However, when compared to legumes, grasses are hardier and last longer; under poor growing conditions grasses outyield legumes; grasses can withstand heavier grazing; legumes can cause bloat, which the grasses cannot; and grasses are just as high in energy and carotene (Vitamin A value).

Hay and silage are the two most common methods of roughage preservation. Hay can be made from grasses or legumes or a combination. Essentially, haymaking is a drying process: moisture content is reduced from 70 to 80 percent down to 20 to 25 percent, at which point the hay is stored. Nutrient losses in haymaking can range from slight to significant (5 to 50 percent), occurring in various ways in varying degrees: (1) Mechanical losses occur when leaves, which are highest in TDN, dry faster than stems and are shattered during mechanical harvest. This problem is particularly significant with the legumes. (2) Oxidative and fermentative losses result from the heating of wet plants. Quick curing will reduce these losses. Carotene content is especially affected. (3) Leaching is caused by a light rain delaying the drying of the crop, resulting in fermentative losses. Heavy rain will leach out the soluble nutrients, causing even greater losses.

Silage is usually made from corn or from hay crops. The advantages of silage are: there is less nutrient loss than with field-cured crops; it saves on concentrates because animals will eat more roughage when fed silage; and it is easily handled by mechanical feeding systems. Corn silage yields more TDN to the acre than does corn grain, and, at a fairly low cost per pound of TDN, provides a high-quality succulent feed year round. (See Appendix II for more on hay and silage.)

Hay Equivalent The term "hay equivalent" is often used to express the forage consumption of livestock, the amount of which is related to body weight. A hay equivalent (H.E.) is 1 pound of hay per 100 pounds of body weight. Generally, cows will consume approximately 2 H.E.s per day of decent quality hay, but may consume as little as 1½ H.E.s if the roughage is poor or as much as 3 H.E.s if the roughage is excellent. A 1,000-pound cow, therefore, will consume between 15 and 30 pounds of hay per day, depending on the quality of the hay. When considering silage, however, a different ratio is used. It has been found that a cow will eat 3 pounds of silage for each pound of hay, so 1 H.E. equals 3 pounds of silage. A ration based on 2 H.E.s per day for a 1,000-pound cow, for example, can mean either 20 pounds of hay or 60 pounds of silage.

Concentrates

Because roughages are generally low in TDN, for many forms of production some additional source of energy is required. Concentrates fulfill this need, as the highly digestible starch of most cereal grains provides a rich supply of energy. Some concentrates are also quite high in protein content; these are the protein supplements, which include soybean meal, cottonseed meal, meat and fish meal, and dried milk products, to name a few. Cereal grains, being high in starch, low in fiber, and high in digestibility, have a TDN range of about 70 to 80 percent. They're also extremely palatable—there's nothing like a bucketful of grain to attract the attention of even the most unruly critter. (Concentrates soothe the savage beast?) Protein levels in the cereals, however, are low (9 to 14 percent), and the protein is of poor quality: enter the protein supplements. Plant protein supplements (oilseed meals) are high in protein (35 to 50 percent) as well as energy. Animal protein supplements (tankage, meat and bone meal, fish meal, etc.) are high in the quality of their protein as well as in quantity.

Livestock rations are first worked around the available roughages, primarily because they usually provide a cheaper source of nutrients than do concentrates. Cereal grains are then added to the ration to make it meet the TDN requirement of the animal. If the protein level is deficient, a protein supplement is added until the requirement is met. Nonruminant rations are also based on the cereals as the cheaper source of TDN; protein supplements, the most expensive portion of the ration, are added only as necessary.

Formulating Rations

The whole idea in feeding livestock is to feed that which is required for the age and production level of the animal as economically as possible. Extensive research has resulted in tables—feeding standards—that serve as guides to the nutrient requirements of an animal at various stages of maturity and production. The standards recommend the amount of dry matter to be fed, digestible protein, TDN, calcium, phosphorous, and carotene requirements. They don't usually specify the vitamins or minerals or the type or quality of protein. Therefore, when developing rations, meeting the feeding standards is not the only consideration.

First, the ration must be suitable in other ways for the class of livestock involved. For example, simply meeting the protein requirement of a pig is not enough; the protein sources must be balanced to provide the high-quality protein required by the pig. Then, the type of feed must be taken into account; it must be palatable to and usable by the species being fed. You can't, for example, feed a chicken on a diet of hay, even though the hay might technically meet the requirements of the chicken. Water content of the feed must also be considered; a cow can eat just so much silage or pasture because her capacity for moisture is limited. For this reason, spring pastures may sometimes require supplementation for livestock; even though the forage's composition may supply the necessary nutrients, the animal cannot eat enough of it to meet its needs because of the water content. Economics, too, is of prime importance; you can't just fill your rations with any old thing, forgetting about cost. Rations must be economically as well as nutritionally sound. Toxicity of certain ingredients can further complicate matters. Cottonseed meal, for example, contains a substance, gossypol, which is toxic at various levels to nonruminants. Cottonseed meal cannot make up more than 10 percent of a swine ration or 5 percent of a poultry ration for this reason.

If all this information makes it sound complicated to formulate your own rations, well, it is, especially for pigs and poultry or for any simple-stomached animal where protein quality must be maintained. For the stockholder with just a few animals, ration formulation is best left to the experts. Many commercial dairies, as well as commercial egg and broiler producers, use commercially mixed feeds. But if you have grains available or if you want to be sure that you are feeding as required, it is a good idea to know how to use the feeding standards. (For techniques in formulating simple, balanced rations, see Appendix I, which includes a table of the proximate analyses of some common feeds.)

However, in most cases you will not need to formulate your own concentrate mixes. Commercial dairy feed, which is also suitable for goats and sheep, is readily available. Also, some already balanced sample grain mixes for swine, sheep, and horses are presented in subsequent chapters.

References

CASSARD, D., and E. M. JUERGENSON. *Approved Practices in Feeds and Feeding.* Danville, Ill.: The Interstate Printers and Publishers, 1971.

MAYNARD, L. A., and J. K. LOOSLI. *Animal Nutrition.* New York: McGraw-Hill, 1969.

MORRISON, F. B. *Feeds and Feeding.* Claremont, Can.: The Morrison Publishing Co., 1961.

CULLISON, A. E. *Feeds and Feeding.* Reston, Virginia: Reston Publishing Company, Inc., 1975.

2
ANIMAL
BREEDING

Variation in Nature

Isn't the variation in nature amazing? First, consider all the different species, the number of them, how they differ from each other—how different a spider is from an elephant, for example—and how they each fill their niche in the environment. Then, there's all the variation within a species, such as different breeds. Chihuahuas and St. Bernards are, incredibly enough, both members of the same species (*Canis familiaris*) and could, actually, be successfully crossed with one another (unlike animals of different species, which under normal circumstances, because of geographical distribution or different chromosome number, cannot be bred together and produce viable offspring). Or take Shetland ponies and Belgian draft horses—again, crossable breeds within the same species (*Equus caballus*), although vastly different in appearance.

Even within a breed, individuals, although similar, are still distinct. Look at two German shepherds. Although certain traits will be uniform enough to tell you that these are both German shepherds, after being introduced you'll find enough difference to make it easy to distinguish Rover from Rambler, even if they're siblings. And the more accustomed you are to looking at the given type, the more you'll pass over the similarities and see only the differences. To a non-cow person, all cows look alike, but to a dairyman, each cow is an individual easily distinguished from her herdmates. The variation is not limited to size and coat color, either; the cows will vary in milk production, calving ease, fertility, feed efficiency, etc.

Two major factors affect variation among individuals in a population: heredity and environment. *Heredity* is the genetic makeup of the animal. *Environment* is a combination of many external forces, such as climate, habitat, feed resources, etc. In domestic animals, environment is greatly influenced by management; a cow giving 10,000 pounds of milk may, in fact, be capable of giving 15,000 pounds if fed and managed differently. Identical twin goat kids reared under different conditions may perform quite differently from each other as adults; the difference would be caused by environment, since as twins their heredity would be the same.

Both heredity and environment contribute to the performance of the animal, and in breeding livestock, one is faced with the knotty problem of separating the contributions of one from the other. Obviously, only differences caused by heredity can be passed on to the next generation, and it is only those differences that can be selected and bred for. It should be equally obvious that it is precisely this genetic variation among individuals that allows for selection and improvement of livestock. The fact that cows within a herd, for example, where environmental differences are minimized, still differ from one another at least partially because of their genetic makeup allows the breeder to select the best ones to reproduce, thereby improving the herd. On a much broader basis, it can be seen that variation among individuals within a species is necessary for the evolution and survival of that species.

The Mechanics of Heredity

Within the nucleus of a cell are a number of rod-shaped bodies known as chromosomes, the number of which varies with each species. Although chromosomes are normally all jumbled up within the nucleus, through various techniques it has been discovered that chromosomes come in pairs—in other words, that there are two of each type of chromosome. Thus, humans, with forty-six chromosomes, have twenty-three pairs, two of each of twenty-three types of chromosome. The chromosomes themselves are composed of genes, which in turn are made up of strands of DNA (deoxyribonucleic acid) that do the actual "programming" of the genetic information.

Genes are considered the basic unit of heredity. A specific gene—say the gene for eye color—is located at a specific spot on the chromosome called a locus. Because chromosomes occur in pairs, it follows that genes, too, occur in pairs. Considering all the members of a breeding population, there may be many types of genes that

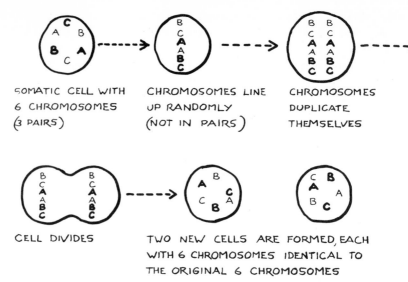

Fig. 2-1. Mitosis

can operate at a given locus. At the eye color locus, for example, within the population you might find a brown type gene, a blue type, a black, etc. Different types of genes at a locus are called alleles. The two alleles that occur in any one individual at the specific locus on the chromosome pair interact to produce the characteristic coded for by that gene—what color eyes the individual will have, in this case.

Before considering interactions of alleles, one should understand how genetic information is passed from generation to generation.

Mitosis and Meiosis

In a living organism, cells are constantly duplicating themselves: dead cells are being replaced, the organism is growing or repairing itself, etc. All the body (somatic) cells contain the same number and types of chromosomes. When a cell divides into two new cells, the chromosomes duplicate themselves and then separate into the two cells by a process known as mitosis. A greatly simplified illustration of mitosis is shown in Fig. 2-1.

In this way, each cell of the body contains the same genetic blueprint as every other cell. What happens, though, with egg and sperm cells, the specialized cells of reproduction? At fertilization, cells from two parents merge to form one new cell. If each parent cell contained a full complement of chromosomes (say forty-six, as in humans), the resultant zygote (new cell) would contain double the number (ninety-six); each generation would have twice as many

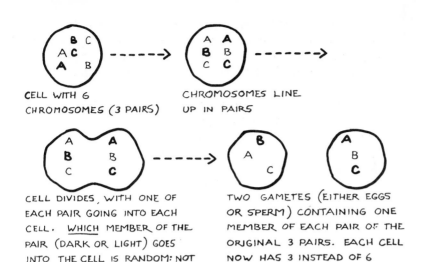

CELL WITH 6
CHROMOSOMES (3 PAIRS)

CHROMOSOMES LINE
UP IN PAIRS

CELL DIVIDES, WITH ONE OF
EACH PAIR GOING INTO EACH
CELL. <u>WHICH MEMBER OF THE
PAIR (DARK OR LIGHT) GOES</u>
INTO THE CELL IS RANDOM: NOT
ALL DARKS GO INTO ONE CELL,
AND ALL LIGHTS INTO THE OTHER

TWO GAMETES (EITHER EGGS
OR SPERM) CONTAINING ONE
MEMBER OF EACH PAIR OF THE
ORIGINAL 3 PAIRS. EACH CELL
NOW HAS 3 INSTEAD OF 6
CHROMOSOMES, SO IS 'HAPLOID'.

Fig. 2-2. Meiosis

chromosomes as the one before. Clearly, this does not happen. In the gametes (sperm and egg cells), a different type of cell division takes place: meiosis. Formerly known as reduction division, meiosis reduces the number of chromosomes from the full number (the diploid number) to half (the haploid number). The union of two haploid gametes, then, forms a new diploid individual. In a human, somatic cells contain forty-six chromosomes, twenty-three pairs, whereas the sperm and egg cells contain only twenty-three, one of each of the original twenty-three pairs. When sperm and egg merge, the new cell gets twenty-three chromosomes (one of each type) from each, and is then complete with the full forty-six chromosomes, two of each of the twenty-three types (Fig. 2-2).

It is important to note here that although chromosomes occur in pairs, the members of each pair are not identical: remember that at any given locus on the chromosome there may be different alleles. So, although a chromosome pair may be illustrated as two A types, the difference between them is denoted by the darker type of one of the pair. When the pairs come together before dividing, they orient themselves randomly. That is, all the darks do not go to one side and into one gamete cell and all the lights to the other and into the other gamete. Instead, the members of each pair are distributed to the gametes randomly. Thus, in the three paired individuals illustrated, the gametes could have ended up with either

A	**B**	**C**		**A**	B	C
A	**B**	C		A	**B**	C
A	B	**C**		A	B	**C**
A	**B**	**C**		A	B	C

It is easy to see that the more chromosomes involved, the more possible combinations there are. The number of chromosome combinations possible is equal to 2^n where n is the number of types of chromosomes involved (the haploid number). In our example, that is 2^3, or 8.

Following two possible gametes produced in this population of six chromosomed individuals through a hypothetical fertilization, we could get

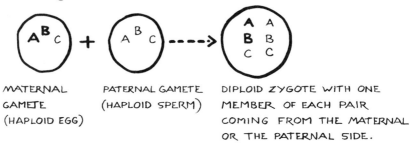

MATERNAL
GAMETE
(HAPLOID EGG)

PATERNAL GAMETE
(HAPLOID SPERM)

DIPLOID ZYGOTE WITH ONE
MEMBER OF EACH PAIR
COMING FROM THE MATERNAL
OR THE PATERNAL SIDE.

The number of possible combinations in the zygote is quite large: $(2^n)^2$ where n is the haploid number. In the three-paired sample organism, the number of possible combinations from the mating of two of these beings would be $(2^3)^2$ or 64. With haploid numbers like 23 in humans, or 30 in cattle, you can see that the possible chromosome combinations in the zygote are enormous, and suddenly, the great amount of variation seen among individuals within a species becomes more understandable.

Now that it is understood where the chromosome pairs come from, we can take a look at how chromosomes and genes can interact.

Gene Interactions

The simplest type of gene interaction is *dominance*. Eye color in humans is a commonly used example of simple dominance. At the eye color locus on the chromosome, two alleles may exist: the gene for brown, designated as B, or the gene for blue, noted as b (following the tradition of the dominant gene being capitalized). In the population, therefore, an individual, on its two homologous (same type) chromosomes, could have a B and a B, a B and b, or a b and b gene. A person carrying the same gene on both chromosomes, that is, a B and B or a b and b, is said to be homozygous for the trait concerned (eye color, in this case); a person with one of each, B and b, is heterozygous. A BB person, having only brown eye genes, will be brown-eyed. The bb homozygote will be blue-eyed. Gene interaction occurs in the heterozygous individual: the Bb person (one brown eye gene, one blue eye gene). The B gene is dominant over the recessive b; the effects of the blue will be covered

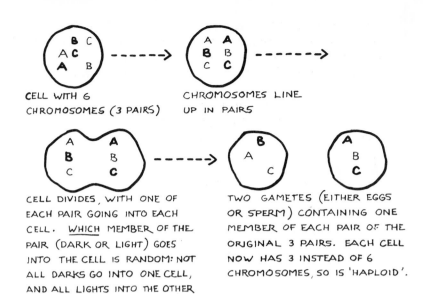

CELL WITH 6
CHROMOSOMES (3 PAIRS)

CHROMOSOMES LINE
UP IN PAIRS

CELL DIVIDES, WITH ONE OF
EACH PAIR GOING INTO EACH
CELL. WHICH MEMBER OF THE
PAIR (DARK OR LIGHT) GOES
INTO THE CELL IS RANDOM: NOT
ALL DARKS GO INTO ONE CELL,
AND ALL LIGHTS INTO THE OTHER

TWO GAMETES (EITHER EGGS
OR SPERM) CONTAINING ONE
MEMBER OF EACH PAIR OF THE
ORIGINAL 3 PAIRS. EACH CELL
NOW HAS 3 INSTEAD OF 6
CHROMOSOMES, SO IS 'HAPLOID'.

Fig. 2-2. Meiosis

chromosomes as the one before. Clearly, this does not happen. In
the gametes (sperm and egg cells), a different type of cell division
takes place: meiosis. Formerly known as reduction division, meiosis
reduces the number of chromosomes from the full number (the
diploid number) to half (the haploid number). The union of two
haploid gametes, then, forms a new diploid individual. In a human,
somatic cells contain forty-six chromosomes, twenty-three pairs,
whereas the sperm and egg cells contain only twenty-three, one of
each of the original twenty-three pairs. When sperm and egg merge,
the new cell gets twenty-three chromosomes (one of each type) from
each, and is then complete with the full forty-six chromosomes, two
of each of the twenty-three types (Fig. 2-2).

It is important to note here that although chromosomes occur
in pairs, the members of each pair are not identical: remember that
at any given locus on the chromosome there may be different alleles.
So, although a chromosome pair may be illustrated as two A types,
the difference between them is denoted by the darker type of one of
the pair. When the pairs come together before dividing, they orient
themselves randomly. That is, all the darks do not go to one side
and into one gamete cell and all the lights to the other and into the
other gamete. Instead, the members of each pair are distributed to
the gametes randomly. Thus, in the three paired individuals illustrated, the gametes could have ended up with either

A	B	C		A	B	C
A	B	C		A	B	C
A	B	C		A	B	C
A	B	C		A	B	C

It is easy to see that the more chromosomes involved, the more possible combinations there are. The number of chromosome combinations possible is equal to 2^n where n is the number of types of chromosomes involved (the haploid number). In our example, that is 2^3, or 8.

Following two possible gametes produced in this population of six chromosomed individuals through a hypothetical fertilization, we could get

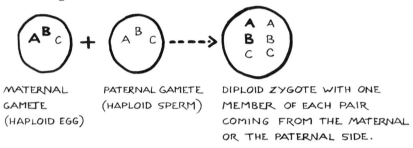

MATERNAL   PATERNAL GAMETE   DIPLOID ZYGOTE WITH ONE
GAMETE   (HAPLOID SPERM)   MEMBER OF EACH PAIR
(HAPLOID EGG)   COMING FROM THE MATERNAL
                                       OR THE PATERNAL SIDE.

The number of possible combinations in the zygote is quite large: $(2^n)^2$ where n is the haploid number. In the three-paired sample organism, the number of possible combinations from the mating of two of these beings would be $(2^3)^2$ or 64. With haploid numbers like 23 in humans, or 30 in cattle, you can see that the possible chromosome combinations in the zygote are enormous, and suddenly, the great amount of variation seen among individuals within a species becomes more understandable.

Now that it is understood where the chromosome pairs come from, we can take a look at how chromosomes and genes can interact.

Gene Interactions

The simplest type of gene interaction is *dominance*. Eye color in humans is a commonly used example of simple dominance. At the eye color locus on the chromosome, two alleles may exist: the gene for brown, designated as B, or the gene for blue, noted as b (following the tradition of the dominant gene being capitalized). In the population, therefore, an individual, on its two homologous (same type) chromosomes, could have a B and a B, a B and b, or a b and b gene. A person carrying the same gene on both chromosomes, that is, a B and B or a b and b, is said to be homozygous for the trait concerned (eye color, in this case); a person with one of each, B and b, is heterozygous. A BB person, having only brown eye genes, will be brown-eyed. The bb homozygote will be blue-eyed. Gene interaction occurs in the heterozygous individual: the Bb person (one brown eye gene, one blue eye gene). The B gene is dominant over the recessive b; the effects of the blue will be covered by the presence of the brown, and the person will be brown-eyed, undistinguishable physically from the homozygously brown-eyed individual, although genetically different. That is, both individuals have the same phenotype (look the same) with different genotypes (genetic makeup). Just by looking at brown-eyed individuals, it is impossible to tell whether they are heterozygous for the trait or homozygous. The blue-eyed phenotype, however, must be a homozygous bb genotype since the b gene is recessive. By looking at the trait in the offspring of a brown-eyed individual, the genotype of the parent may be determined.

The following three examples illustrate various crosses involving homozygous and heterozygous individuals. (Study the key first.)

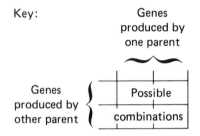

Example 1: Cross a homozygous brown-eyed individual (produces only B gametes) with a blue-eyed individual (produces only b gametes). The homozygous recessive (bb) can give *only* a b to the new individual; the homozygous dominant (BB) can give *only* a B to the new individual. The possible combinations are:

	b	b
B	Bb	Bb
B	Bb	Bb

Result: Offspring are *all* Bb, having received a B from one parent and a b from the other. All are brown-eyed.

Example 2: Cross a heterozygous brown-eyed individual (Bb) with a homozygous recessive (blue-eyed) individual. The heterozygote will produce both B gametes and b gametes in equal numbers. The homozygote will produce only b gametes.

	b	b
B	Bb	Bb
b	bb	bb

Result: Crossing a heterozygote with a homozygote results in both heterozygous and homozygous offspring in a 1:1 ratio: half the offspring are brown-eyed, half are blue-eyed.

Comment: If a brown-eyed person and a blue-eyed person have a blue-eyed child, the brown-eyed person *must* have been heterozygous (Bb) rather than homozygous (BB) in order to be able to donate one of the two b genes found in the blue-eyed offspring. If, however, the offspring is brown-eyed, the brown-eyed parent may have been *either* homozygous (BB) or heterozygous (Bb), since in either case, a dominant B gene could have been donated resulting in the brown-eyed offspring. The genotype of the parent showing the dominant characteristic (brown eyes) is proven *only* if and when the recessive trait (blue eyes) comes out in the offspring. If the recessive trait shows up in even one offspring, the genotype of the parent is proven to be heterozygous (Bb). If the recessive trait *never* shows up, even with hundreds of offspring (as in cattle), the genotype of the parent, while suspected to be homozygous, is not proven to be so, since chance alone may dictate that result from a heterozygous individual.

Example 3: Cross a heterozygote with a heterozygote.

	B	b
B	BB	Bb
b	Bb	bb

Result: 75% of the offspring are brown-eyed, 25% are blue-eyed. This is a 3:1 *phenotypic* ratio. However, two of the brown-eyed offspring are heterozygous (Bb) and one is homozygous (BB). This is a *genotypic* ratio of 1:2:1, or one homozygous dominant, two heterozygous, and one homozygous recessive.

Incomplete dominance, or incomplete inheritance, can also occur between alleles. Consider, for example, roan coloring (mixture of white and colored hairs) in some cattle and horses. In the total population, there is the gene for white, W, and nonwhite, w. An individual can be WW, Ww, or ww. WW is white, ww is nonwhite, or colored. Ww, however, where there is no dominance, will show both the white and the colored hairs, and be roan.
Other types of gene interactions include *epistasis,* where the phenotypic effect of one gene is masked by the presence of an entirely different gene (this interaction is nonallelic; dominance, which this resembles, occurs between alleles); *collaboration,* where two different genes interact to produce one trait that neither gene

could produce alone; *modifiers,* where many genes exert effects on a trait. Environmental effects may influence the phenotypic expression of a trait, or mutations might occur, some accident during gene duplication giving rise to a new gene, producing a new effect on a trait. More complicated interactions occur when there are more than two alleles for a given locus, for example with A, B, and O blood types in humans.

Consideration of the chromosome complement as a whole reveals still more areas for producing the variation seen among individuals. If genes for different traits are located on nonhomologous chromosomes, they will be inherited completely independently of one another. If, on the other hand, the two genes for the two traits are located on the same chromosome, they will more often be inherited together. Genes for eye color and height are located on nonhomologous chromosomes and so are inherited independently. Genes for eye color and hair color, however, are located at different loci but on the same chromosome and are said to be *linked.*

The distance between loci on a chromosome determines how closely linked the two traits are. If genes for a trait are closely linked, it may be very difficult to separate them (as might be desired if trying to select for one trait and against another in animal breeding). With less closely linked genes, some exchanges of genetic material could occur during gamete formation through *crossing over* (Fig. 2-3).

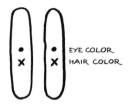

CHROMOSOME
TYPE I

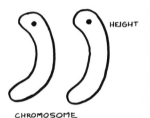

CHROMOSOME
TYPE C

Fig. 2-3. Crossing over Between Chromosomes

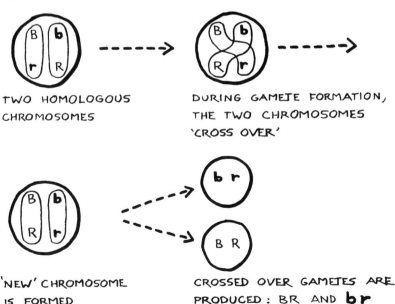

TWO HOMOLOGOUS
CHROMOSOMES

DURING GAMETE FORMATION,
THE TWO CHROMOSOMES
'CROSS OVER'

'NEW' CHROMOSOME
IS FORMED

CROSSED OVER GAMETES ARE
PRODUCED : BR AND **br**
INSTEAD OF THE ORIGINAL
B**r** AND **b**R

Sex Determination

So far it's been stated (repeatedly) that chromosomes occur in pairs, with each member of the pair being of the same type. This statement is not strictly true: males and females have different chromosome complements, where one or the other sex has a chromosome pair that is not of the same type. In mammals, the female has a pair of X chromosomes, and the male has an X chromosome paired with a much smaller Y chromosome. Human females, therefore, have twenty-three chromosome pairs, one of which is XX, whereas males have twenty-two pairs, plus one set of an X and a Y. During gamete formation (meiosis), females produce only X-bearing cells (ova), whereas males produce both X-bearing and Y-bearing cells (sperm). Since the female can contribute only an X chromosome to the zygote, it is the male who determines the sex of the offspring: if a Y sperm fertilizes the egg, the zygote is XY, or male; if an X sperm wins out, the zygote is XX, or female. Work is underway on separating X- and Y-bearing sperm—if no one else, cattle breeders would love to choose the sex of future calves.

Apparently there are few genes on the Y chromosome, but whatever genes there are obviously control traits in the male only. Genes found only on the Y chromosome are termed holandric.

On the other hand, there are many genes found on the X chromosome not found on the Y; these are sex-linked genes. Since a female has two X genes and a male has only one, recessive sex-linked genes cannot be masked in the male, and therefore, the recessive trait will show up much more easily in the male than in the female. Two such recessives in humans which normally only show up in males are color blindness and hemophilia.

The following three examples show the various possibilities for inheriting sex-linked color blindness. (Study the key first.)

⊗ COLOR-BLIND GENE
x NORMAL GENE
y NORMAL GENE
x x NORMAL FEMALE
x⊗ CARRIER FEMALE (NOT COLOR-BLIND SINCE x IS DOMINANT)
x y NORMAL MALE
⊗y COLOR-BLIND MALE (⊗ RECESSIVE COMES OUT SINCE NOT
     MASKED BY A DOMINANT x GENE)

Example 1: Non-color-blind male (XY) and carrier female (X⊗).

	x	y
x	x x	x y
⊗	⊗x	⊗y

Result: 25% normal females
25% carrier females
25% normal males
25% color-blind males

Example 2: Color-blind male (ⓍY) and normal female (XX).

	Ⓧ	Y
X	XⓍ	XY
X	XⓍ	XY

Result: All females are carriers.
All males are normal.

Example 3: Color-blind male and carrier female.

	Ⓧ	Y
X	XⓍ	XY
Ⓧ	ⓍⓍ	ⓍY

Result: 25% carrier females
25% color-blind females (the two recessive, Ⓧ, one donated from each parent, makes the offspring homozygous and showing the trait).
25% normal males
25% color-blind males

The preceding facts apply only to mammals; the situation is slightly different in poultry. Males in this case have two identical chromosomes, XX, and females have only one sex chromosome, also an X. The male, then, produces only X sperm, whereas the ova produced by the female can be either X or zip, no sex chromosome at all. The female, now, determines the sex. If an X sperm fertilizes an X egg, the chick is male; if the X sperm fertilizes an "o" egg, the chick is female.

Population Genetics

The inheritance patterns mentioned so far have all been fairly simple, concerning traits influenced by one or two genes. But when dealing with livestock, although it's all very well to understand coat color inheritance, there are a myriad of more important, economically significant traits, which have a much more complex inheritance. Variation in milk production, for example, which is certainly an economically important trait, is extremely complex and may be influenced by as many as two-hundred pairs of genes. Also, the animal breeder is interested in herds, or whole populations, rather than individuals, so thousands of genes from thousands of animals are involved. In such a situation, the simple percentages and ratios mentioned earlier don't apply. Rather, phenotypic variation is examined as a continuous distribution including the entire population instead of individuals. Look, for example, at the graph showing height in humans. You don't see just a few different heights but a whole range, with the bulk of the population centering around the "average" height, with fewer and fewer people representing the heights found as you move away from the average and out toward the extremes (Fig. 2-4). So, animal breeders use population genetics, where the effects of great numbers of genes considered together are studied on the average of the population as a whole.

Heritability

It has already been stated that differences seen in animal performance can be attributed to both genetic as well as environmental factors. Environmental factors are the easiest to control and change in order to improve the herd. Genetic improvement, however, is a slow process in most domestic livestock. (The shorter the interval from one generation to the next, the faster progress is possible. In rabbits, or even swine, for example, there is the potential for faster genetic improvement than in cattle or horses.) The

Fig. 2-4. Normal Distribution Curves

68% of the population is found between the mean and ± one standard deviation

95% of the population is between ± two standard deviations

99.7% is between ± three standard deviations

68% of the population (women) is between 5'5 and 5'9

95% of the population is between 5'3 and 5'11

99.7% of the population is between 5'1 and 6'1

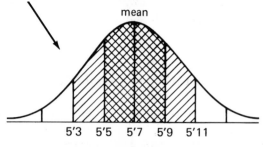

differences in animal performance, therefore, must be separated into those differences that are the result of environment and those that are genetic.

Population geneticists have researched this question in many traits and discovered that the proportion of variation found in the trait that is attributable to genetics and not to environment varies with the trait. Heritability is the term used to describe this proportion, or percent. For example, fleece weight in sheep has a heritability estimate of .33, which means that 33 percent of the variation seen in fleece weights is caused by genetic factors and 66 percent by environmental differences. Litter size in swine has a fairly low heritability of .17; that is, only 17 percent of the variation seen in litter sizes is attributable to the genetic makeup of the individuals involved; the rest is environmental (such as management at farrowing time).

This does not mean that one should select only for those traits with high heritability, even though, if all other things are equal, more genetic progress could be made by doing so. The economic value of the traits must be taken into account (milk production with its low heritability versus coat color with its high heritability, for example). Some heritability estimates are given in Table 2-1.

Selection and Mating Systems

Since genes may influence more than one characteristic, selecting for one trait usually affects other traits as well; these other effects and relationships should be kept in mind when selecting breeding animals. Selecting for rate of gain in swine, for example, results in a genetic improvement in the efficiency of feed conversion, but it also will result in increased fatness. High milk production in dairy cows is negatively correlated with the percent of milk fat. In beef cattle, selection for daily gain increases efficiency and mature size, but selection for mature size does not significantly increase efficiency.

Determining which are the best animals to breed can be accomplished in a few ways. You could just look at the animal and base your decision on phenotype; depending on the trait, you could be more or less successful. A trait such as face covering in sheep responds well to phenotypic selection. Many traits, however, are more complex and may be hidden (carcass characteristics) or sex-linked (how do you choose a bull for milk production?). For these traits, progeny testing is used. Records from many, many offspring from selected mates are compared, and evaluations of the breeding animals are then based on the performance of their offspring. Progeny testing is the basis for sire evaluation for artificial insemination in cattle. Pedigrees are also useful in selection, but often too much emphasis is placed on a long-removed famous ancestor who, by the

Table 2-1. Heritability Estimates for Some Traits in Various
Species of Livestock

TRAIT		HERITABILITY (PERCENT)
Beef cattle	birth weight	35-40
	weaning weight	25-30
	calving interval	0-15
	yearling weight	40-45
Dairy cattle	milk yield	20-30
	percent fat	40-60
	mature weight	60
	feed efficiency	30-40
	mastitis resistance	10-30
Sheep	birth weight	30
	face covering	45-55
	weaning weight	10-40
	mature weight	40-60
	number of lambs born	30
	wool weight	30-60
Swine	number of pigs farrowed	5-15
	number of pigs weaned	5-15
	weaning weight	10-20
	6-month pig weight	20-30
	feed efficiency	30-50
	backfat thickness	40-60
	percent carcass in ham	40-60
	number of nipples	20-40
Chickens	egg production, light breeds	10-20
	egg production, heavy breeds	25-40
	egg size	35-90
	mortality (first year)	5-10

time it gets down to your animal, is making very little genetic contribution.

Once the breeding animals are selected, the question of mating system arises. At this point, a new twist to the heritability and breeding situation must be added. Although it might seem that lowly heritable traits are a waste of time to select for, it happens that a lot of progress can be made on these traits by using cross-breeding because of the effects of *heterosis*. Heterosis, also known as hybrid vigor, exists when, for a given trait, the performance of the crossbred individual is superior to the average of the parents. Interestingly, traits that are low in heritability are high in heterosis. Litter size in swine, egg production in hens, and fertility, calving percent, and weaning weight in beef cattle are traits that, although low in heritability, respond excellently to cross-breeding due to heterosis. Research has shown that in commercial cow-calf herds, where the owner raises straightbred calves from straightbred cows, a switch to crossbred calves from crossbred cows could increase the pounds of calf produced per one hundred cows bred by 19 percent. (Once the crossbred herd was established, however, further improvement would depend on the selection of superior individuals within it.) One can easily see that the existence of heterosis is of prime importance when dealing with economically important traits that are low in heritability.

Two major types of mating systems are inbreeding and out-breeding, with variations. *Inbreeding* is the mating of closely related animals (siblings, son/dam, etc.) and tends to make the offspring more homozygous, thereby maintaining genetic uniformity and increasing the breeding predictability. Inbreeding can reduce the performance of individuals, because of the increased homozygosity, but the use of inbred strains for crossbreeding purposes has been very successful. In *linebreeding,* a milder form of inbreeding, a close relationship is maintained to some superior animal or animals, but the offspring are not highly inbred. *Outbreeding* is the mating of individuals to produce hybrids, which of course results in increased heterozygosity. Hybrid vigor is increased as the relationship between the animals crossed is decreased. Crosses between breeds (*crossbreeding*) show greater heterosis than those between families within a breed (*top-crossing*). If the individuals are from inbred strains of different breeds (*incross bred*), the hybrid vigor of the offspring is even greater.

For the Smallholder

All this talk of inheritance, selection, and breeding systems based on herds may be very interesting (or may be deadly dull), but how does it relate to the person with one cow, a couple of goats, or a pig?

The first thing that a smallholder should realize is that the environment in which the animal functions plays a major role in the performance of that animal; one simply cannot write off poor performance as "bad blood." A good deal of thought and attention to management will go far in achieving top production from your stock.

Then, anyone keeping livestock as more than just pets must be concerned with the economics of the situation. An understanding of inheritance patterns, selection, and mating systems will help in making more informed, intelligent decisions about how to handle your stock when breeding time comes around. Whether you are breeding one sow or fifty, you should be concerned with the performance of the sow (e.g., litter size) and of the piglets (weaning weights, carcass traits, etc.) You should ask yourself what you are aiming for in the piglets (e.g., a bacon type or a ham type) and select a boar accordingly rather than just breeding your sow to any old boar around. The same is true when selecting a bull for your cow, a stallion for your mare, a billy for your nanny, and so on. Ask yourself what you want in the offspring, and choose the sire intelligently. Be aware of the relationship between traits—such as the negative correlation between milk yield and butterfat—and the heritability question—some of the characteristics you may be interested in may not be very selectable. Chickens, goats, beef cattle, horses, what have you, it pays to breed wisely and/or to choose to buy animals that have been bred wisely.

The smallholder should also recognize the potential for fun in playing with the breeding decisions for their animals. Breeds of animals as they exist today were largely developed by humans, after all. Differences that existed naturally among individuals in a species were capitalized on for specific purposes through careful selection and breeding. New breeds can be developed by anyone with the time, patience, skill, and money to do so; such is the case with many new beef breeds, for example the beefalo, a cross between a beef breed and a buffalo. So why not indulge your fancy a bit, and if you can afford to do so, try some crossbreeding? Who knows, maybe you'll end up with a fabulous new breed.

References

CROW, J. F. *Genetics Notes,* 7th ed. Minneapolis, Minn.: Burgess Publishing Co., 1976.

HUTT, F. B. *Animal Genetics.* New York: The Ronald Press Co., 1964.

LERNER, I. *The Genetic Basis of Selection.* London: John Wiley and Sons, Inc., 1964.

II.
DAIRY
ANIMALS

3. DAIRY COWS
4. DAIRY GOATS

Introduction

Acquisition of a dairy animal is the ultimate step in building up the homestead livestock repertoire. She dominates all else: *she* determines the chores schedules, *she* determines how long chores will take, and *she* looms largest in annual calculations. She deserves to reign supreme, however, because she also contributes the most. What a joy it is to sail past the dairy case in the supermarket. Milk, butter, yogurt, cheese, ice cream—it all means nothing because you're producing your own. (Although I must admit that I derive a certain malicious satisfaction from checking out the dairy prices.) Most smallholders can produce milk for about 10 cents a quart (whether goat or cow milk), covering out-of-pocket expenses (feed, hay, miscellaneous equipment), and 25 cents a quart counting everything (labor, capital equipment, etc.). Once the milk is converted into other products, the savings are even greater; remember, the more processing a raw product undergoes, the more middleman costs you save by doing it yourself. Yogurt, for example, costs the homesteader nothing but the milk and five minutes of labor. A quart of yogurt, which costs $1.50 in the store, costs the homesteader, at most, about 50 cents. The same sort of comparison holds for other dairy products. However, although there's not much doubt about the economics of keeping a dairy animal, there's also not much doubt about the work involved. And there *is* work—lots of it.

Dairy animals must be milked on a set schedule. Whether you decide to milk at 7:00 A.M. and 7:00 P.M. (a twelve-hour interval) or

it—every day, morning and evening, ten months a year. You may
not feel like it, especially on those cold, dark, winter mornings, but
there's no staying in bed that extra hour. Milking schedule is the
big, obvious "Dairy Animal Tie Down," but what about the less
obvious ones?

One hidden item of work is dealing with the milk. If you milk
the animal, use enough milk for your coffee, and then watch the
rest go sour, what's the point? To be economical, you have to keep
up with your milk supply. Another item is the breeding/calving or
kidding cycle. To keep the milk coming, you have to breed the
animal, which ostensibly results in an offspring of some sort. The
offspring has (or have) to be dealt with. Management is another
major issue. Dairy animals are (or should be) highly productive,
and therefore they need much more skilled attention than do other
livestock, particularly in nutrition. You can get away with sloppy
management a lot less with a dairy animal than with most other
beasts.

All this discussion is not meant to scare you off; it's meant to
make you think, and thinking ahead is the most important part of
taking the ultimate step. What are your dairy needs? How much
time do you have? How will her schedule fit in with yours? How
much do you already know about dairy animals, and how can you
learn more (and there is always more to learn)?

For my part, I think dairy animals are great and are a valuable
asset to your livestock complement. If you think you can do it, get
as much general background as you can, then do it. The rewards
are enormous—and I'm not talking just about dairy products but
also about the satisfaction and pleasure you get from your queen of
the barnyard.

Cows or Goats?

Get a bunch of homesteaders together and inevitably the great
debate will be raging: dairy animal—great thing to have around—
but which, cow or goat? The arguments are strong on both sides,
especially since goat people tend to be well-armed enthusiasts of
their dairy underdog against the dairy industry's heavy favorite. I
have seen (in print) and heard (in words) an awful lot of mud slung
from both sides: amazingly erroneous and unfounded statements
made by goat people about cows (they're hard to manage) and by
cow people about goats (they eat tin cans). People do seem to get
passionate when defending their milk supply.

Goats are definitely in order when land and space are limited.
A goat needs less space than a cow for housing, pasture, and hay
storage. A person with a few acres and a small shed would probably

be better off with two goats than with a cow. Certainly a cow doesn't have to be grazed, but it is much more economical if she can be. A goat, on the other hand, because she requires less total feed, can still be economical when fed harvested feeds. Winter storage space for hay is also a consideration; less storage space is needed for goats because of their smaller requirements. If you do have a small grazing area available, but it is weedy and covered with little shrubs, a goat can get some feed value out of it, whereas a cow in the same position would just stand around soaking up the sun. Manure handling is another factor limited by land resources. Goat manure is not at all bulky; it consists of nice, dry little pellets. Cow manure, though—well, we all know what cow manure is like. It's not so difficult to deal with, but only if you've got the land (or neighbors' gardens) to spread it on. Of course, these advantages diminish as one keeps more and more goats.

One of the big arguments often used in favor of goats is that they're small and easy to handle. I won't argue that they're small— it *is* an advantage that they don't break your toe when they step on you. But having handled both cows and goats, I certainly would not agree that goats are easier to handle. They're lighter, sure, but they're also smarter, and I am convinced that ease of handling is negatively correlated with intelligence. Cows are content to stand around and watch the grass be green. If you want them to move, they move. Goats, however, are clever and are always thinking of fun things to do, like "let's jump over the fence today" or "let's go out and eat the neighbor's rosebushes." Not that this intelligence doesn't work in favor of goats, too—they have an awful lot of personality, and are incredibly easy to get attached to. But it does, to my mind, make them a little more difficult to manage.

Quantity of milk is cited as another plus for dairy goats. An average good goat will produce about 1,800 pounds of milk over a ten-month lactation, or about three quarts a day. Three quarts a day is a nice, moderate amount of milk, and if you stick to that, great. But the catch is that you don't usually have just one goat. First of all, goats get too lonely alone and need the company of another of their kind. Second, most goat owners find that it is necessary to milk more than one goat once they start making dairy products; milking several goats means dealing with the same quantity of milk as put out by a low-producing cow. Besides (and here's one of the biggies for cows), with cow's milk you're really getting two products: milk and cream. Goat's milk is naturally homogenized, that is, the fat globules are smaller and stay in suspension. This characteristic makes goat's milk a little easier to digest (it's especially good for people with ulcers), but it necessitates a cream separator to extract the cream. Take a gallon of cow's milk, though, let it stand for ten to twelve hours, and presto! Skim off a quart of cream with a plain old ladle, and you've got three quarts of skim milk left—simple and easy. And cream, of course, is the foundation for butter, ice cream, cream cheeses, and those delectable whipped cream delights. Add to this the fact that for the same amount of milk as produced by a cow (and without the easily separated cream) you'd have to handle anywhere from four to eight goats, and you end up with a high labor input for goats when compared to cows on a gallon-for-gallon basis. The labor issue is one that greatly hampers the commercial use of goats; with the smaller milk-producing unit, more animals must be handled per volume of milk. In addition to the increased labor cost of milking several animals in place of one, there is also the increased cost of housing, bedding, breeding, and other management-related requirements for several instead of one. In areas where labor and overhead are cheap, the high labor cost of goats may, in fact, be an advantage, but where time is bought at a premium price, the labor factor is a serious drawback.

Management considerations include land, fencing, and breeding. If you *do* have some land available and decent pasture, a cow will make better use of it than a goat, and more easily. A cow can be contained by any old fence—one strand of electric wire will work fine. Goats, however, need a minimum of a four-foot high fence if woven wire (expensive and difficult to install), or at least two strands of electric (which doesn't always work). The most avid of goat lovers will admit that fencing is a hassle, although as stated before, goats do not have to have pasture but can do well with a small exercise yard.

I have seen it written, in several places, that "goats are so easy to breed—all you have to do is put the doe in the back of your car, take her to the buck, and it's done." The implication was that, after all, you couldn't stick a cow in the back seat of your car. This is undeniably true; I won't even offer a mild protest. But all the cow breeder has to do is pick up the telephone, call the artificial insemination technician, who comes right over, pay the six or eight dollars, and the cow is bred. You can't get much easier than that, aside from the fact that buck stud fees are generally higher than comparable A.I. sire fees.

Then, of course, does give birth to goat kids, whereas cows have a tendancy to drop calves. Economic reality is such that unless you've got great bloodlines in your goats, a bull calf is worth a lot more than a buck kid, and is much easier to sell. In this country beef still appears to be more popular than chevron.

Contrary to what most goat people think, cows and goats use about the same amount of feed to produce the same amount of milk. Goats and cows both eat about two pounds of hay per one hundred pounds of body weight. Goat grain levels are one pound of grain per four pounds of milk, the same as for the larger dairy breeds. (This equation assumes a supply of good feed and roughage. In areas where roughages are very poor, such as in desert areas, goats will survive where cows would die.) Given that the input/output ratios are the same, one should be wary of paying proportionally more for a goat than a cow: in many areas, goats are actually more expensive than cows. In the Northeast, for example, in 1978 a good, moderate-to low-producing family cow cost around $250, but it was difficult to find good goats for less than $85 to $100 each. To get the same amount of milk as put out by that $250 cow, four good goats were needed, at a total cost of $340 to $400. There is one big advantage, though, of having several animals instead of one: it spreads out the cost of possible losses.

Something should be mentioned about the famous "goaty flavor" attributed to goat's milk. Many goat people claim that this flavor comes from poorly kept goats, and that milk from clean goats is indistinguishable from cow's milk. However, I once tasted milk from two does housed together; one tasted goaty, and the other tasted like cow's milk. You got me. It is known that even though goat's milk comes from the goat cleaner (in terms of bacteria) than cow's milk comes from a cow, it picks up bacteria more easily once it's out, which would seem to contribute to the prevalence of off-flavors. If this is the case, the situation can clearly be kept under control with proper sanitation, but it does make the handling of goat's milk a little more difficult.

A last point to be made in favor of goats: goat people are just that. They tend to band together, be very sociable, exchange a lot of information, and generally form an enthusiastic community. Getting goats is like joining a great club, and that can be even more fun than the goats themselves.

Anyhow, what does this all boil down to when it's time to get a dairy animal? Most likely, after you've talked yourself this way and that, weighing the pros and cons of each, all resolution will go out the window when you see a cute, nibbly doe with a fawnlike kid by her side; or perhaps you'll be taken in by a pair of big brown eyes and a gentle moo. It may not be the best way to make the decision, but either way, it'll be loads of fun.

3
DAIRY COWS

The Family Cow

A dairy cow has to be one of the most placidly pleasant animals to have around—which is not to say that there isn't such a thing as a rambunctious cow; on the contrary, I've known a few personally. One cow in particular I remember quite clearly—who wouldn't? An average-sized Holstein (not small, in other words), this cow could put a thoroughbred hunter to shame, as she sailed over fences, hedges, and ditches with nary a second thought. Quite a sight to see, and quite a cow to catch. But the usual cow, especially one who has been a family cow all her life, is a docile, gentle creature who enjoys attention (scratching her around the ears and under the chin will endear you to your cow forever), is happy to be milked, and generally aims to please. They are creatures of habit; teach them a routine, and they will follow it faithfully—as long as it remains *exactly* the same. They *are* big animals, though—even the small breeds—so a certain amount of firmness is required in dealing with them; a slap on the rump will usually do the trick.

No matter how nice her personality may be, however, it is not, after all, why you would keep a cow. Although milk is reason number one, equally important to many family cow owners are the by-products of cream and calves. Cream, of course, especially in the quantities produced by a Jersey or Guernsey, is an extremely valuable product, and unlike the cream in goat's milk, is quite easy to collect. Dairy calves are also valuable, whether sold right away or raised for veal or beef.

Easy to handle, easy to manage, excellent utilizer of pasture, efficient milk producer, the dairy cow has a place with any family needing her bounty.

Breeds

There are five major breeds of dairy cattle in the United States, and certainly anyone contemplating ownership of a dairy cow should be familiar with the characteristics of each.

Holstein–Friesians (called Holsteins here and Friesians in Europe) are the big black and white beasts you see all over the countryside. They are the largest of the breeds (1,500 pounds), the heaviest milkers (about 5½ gallons per day), the lowest in milk fat, and the most popular commercial herd animal. They are gentle animals (even if they are huge) and easy to handle. The good-sized calves (95 pounds) and their rapid rate of growth make Holstein calves and surplus animals quite usable for veal and beef. Although usually black and white, you do run across all black, all white, or even red and white Holsteins. (The red and white color is the result of a recessive gene.) Holsteins originated in Holland, surrounded by lush pastures, and so never needed to be selected for superior grazing ability. As a result, they are not as hardy foragers as are other

Holstein cow, "Beecher Arlinda Ellen."
Courtesy: Holstein-Friesian Association of America.

breeds. Generally, Holsteins are not the best of family cows. Their size requires a lot of feed, and although the feed is efficiently converted into a large quantity of milk, it is usually more milk than most families want. Also, many family cow owners place a high value on the quantity of cream, an item for which the Holstein is definitely not noted. For several families sharing a cow, though, or for extensive cheesemaking, the Holstein is quite an efficient milk supplier.

The Brown Swiss is next in size (1,400 pounds) and in milk production (five gallons per day) to the Holstein. Not surprisingly, Brown Swiss are brown (of varying shades) and originated in Switzerland. The quality of Swiss grazing—rough—produced the excellent grazing ability found in this breed. Brown Swiss cattle also do well for beef and veal, since they, too, have large calves (90 pounds) and large body size. The main drawback for the family is that Brown Swiss are fairly low in butterfat, too. A particular drawback for the beginner is that they also have quite a reputation for stubbornness (but not meanness; sometimes they just don't like to move).

Ayrshires are getting down in size a bit (1,200 pounds) and in milk production (4½ gallons per day) from the other two breeds. Ayrshires are anywhere from red to mahogany with white and are distinguished by large, upturned horns and excellent udders. Originating in Scotland, Ayrshires are very hardy and excellent foragers. Their milk-fat level is about the same as that of Brown Swiss. Although Ayrshire lovers will defend them to the death, they are

Brown Swiss cow.
Courtesy: Brown Swiss Cattle Breeders' Association.

Ayrshire cow, "Leete Farms Betty's Ida," holds Ayrshire record of
37,170 pounds of milk (2X-305 days).
Courtesy: Ayrshire Breeders' Association.

considered to have the most nervous temperament among the dairy
breeds.

These three breeds are pretty easily distinguishable from each
other, and it's easy even for the neophyte to tell them apart from
Jerseys or Guernseys. But most new cow people have a hard time
telling Jerseys and Guernseys apart from each other. You can narrow it down to one of those two—but which? They're approximately
the same size (1,000 pounds), although frequently Jerseys are
smaller. They're both considered fawn-colored—a kind of golden
brown. The key to telling them apart is the presence (or absence)
of black and white—black on Jerseys; white on Guernseys. Guernseys usually have clearly defined white marking; Jerseys may have
some white on them, but nowhere near as commonly as Guernseys.
Jerseys often have some black: black switch (the tuft at the end of
the tail), black around the eyes, or black on the muzzle. Guernseys,
on the other hand, are usually yellowish pink around the eyes and
may have a pink, rather than a black, nose. Also, once all cows stop
looking alike to you, you'll quickly be able to spot the typical "dished
face" of the Jersey cow.

Both Jerseys and Guernseys make excellent—and the most
common—family cows. They're small, good natured, produce moderate amounts of milk (3½ gallons per day with Guernseys, the
heavier producer of the two) and lots of cream (Jerseys win out

here). They both have small calves (Jersey calves are the tiniest, and cutest, things you've ever seen) and so are not considered too good for veal. They're also not considered good for beef because their fat is tinted yellow rather than white; however, this is purely a commercial marketing question and means nothing to the person producing meat for home consumption. Jerseys are better foragers than Guernseys and are more heat tolerant. Jerseys also have an extra little bonus in being notoriously easy calvers.

Some mention should be made of the dual-purpose animal, one considered good for both milk and meat production but not superior in either trait to the straight beef or dairy breeds. In the United States, the Milking Shorthorn is the most prevalent of the dual-purpose breeds. The Devon, Red Polled, Scottish Highland, and Simmental can also be considered dual-purpose animals, the first two being more "milky," the last two more "beefy." Many small stockholders like the idea of a dual-purpose animal. However, if your main interest in keeping a cow is milk, and the beef is an offshoot, you'd do better sticking to a dairy breed and crossing her to a beef breed to produce more beefy-type offspring to raise for meat.

Selecting a Dairy Cow

Shopping for your first dairy cow should be an exciting, and I hope, not too terrifying experience. If you're lucky enough to have a friend experienced in this sort of thing, drag her along. If not (even if so), going off with some ideas in your head about what you're looking for will make the whole event much more worthwhile. Notice, by the way, I said "shopping for," not "buying," your first dairy cow. Some feeble attempt should be made not to fall for the first pretty face you see but to try to be objective, shop around, and keep certain points in mind.

The first point to keep in mind is that farmers have suffered for years from bad public relations: slow, dull-witted, basically dumb, trying to take the city slicker for a ride, then heehawing over the joke back in the haystack. Wrong. Most of the people you talk to about their animals will be friendly, helpful, certainly knowledgeable, and primarily fair. If you ask sensible questions, most people, rather than taking advantage of your ignorance, will open right up and give lots of helpful hints. So don't go pretending you've been in the cattle business for years if you're not quite sure how many teats a cow has. But also don't expect to walk off with a bargain cow by tricking the dumb farmer; most people know what their animals are worth.

Presumably you have already figured out which breed of dairy cattle most appeals to you and best suits your needs. There are

Guernsey cow, "Cleverlands Nancy Judy Jean," the first Guernsey cow to make three records over 30,000 pounds of milk, and the youngest Guernsey to make over 100,000 pounds of milk in a lifetime.
Courtesy: American Guernsey Cattle Club.

three main places to seek out your cow: in an existing commercial herd, from someone selling a family cow, or at an auction. Newspaper ads, notices at feed stores, and conversations with dairy farmers can put you in touch with sellers. At this point, though, I'd say stay away from auctions unless you're very experienced with dairy cattle. Remember that the cow is being auctioned off for some reason, and determining that reason and judging its seriousness can be a tricky thing, even for a veteran dairyman.

Finding a family cow for sale is probably the safest bet. The advantage of a family cow over one from a commercial herd is that it already will be accustomed to hand milking, a big plus for the novice milker. Most cows will adjust from machine milking, but it will be more trying for all concerned if neither of you knows quite what's going on. This is also the reason for the first-time milking person to shy away from the first-time milking cow (a first-calf heifer). First-calf heifers can be pretty frisky beasts, and one thing you don't want when you're learning to milk is a 1,000 pound frisky animal to learn on.

Along these lines, a family cow also has the advantage of having been handled a lot, probably even by children, and so is likely to be quite gentle. A good first question to ask the seller, though, is "Why are you selling her?" A commercial herd may be

Jersey cow, "Generators Topsy," classified Excellent with 96 points
and a top production record of 26,740 pounds of milk and 1,168
pounds of butterfat as a seven-year-old.
Courtesy: American Jersey Cattle Club.

selling a cow because of low production—not necessarily a drawback
for the family cow owner at all, and possibly even a plus. Or the
cow may have had breeding or calving problems, which most people
aren't too eager to tell about, and definitely a drawback for any cow
owner. With any cow, there may be disposition problems (also information not readily volunteered), but your own handling of the
cow and the owner's handling should give you clues. Look at the
animal's pen or fencing: a six-foot high chain link fence for one
heifer might tell you something about her disposition. Then again,
when someone's selling a family cow, very often it's simply because
they got tired of milking her or are moving away, etc.

Okay, you're staring at the beast; what do you see? What do
you look for? First, look at her overall appearance: is her coat good;
are her eyes bright and her eyes and nose clear; is she alert looking
(for a cow); is she sound on her feet (note the way she walks and
stands); does she have any hideous, obvious wounds, especially on
her udder or legs; is she coughing or sneezing; is she being shown
in the dark? Basically, does she appear healthy?

Next, look her over in terms of general body type and what is
called dairy character. All parts of her body should seem in proportion and smoothly put together. Her back should be straight and
strong looking; and her rump, long, wide, and nearly level. Her

forelegs should be straight and wide apart; her hind legs straight from the rear view, and when viewed from the side, nearly perpendicular from hock to pastern. Her neck should be long and lean and clean cut in the throat. She should be sharp over the withers, with wide, well-sprung ribs and loose, pliable skin. Her body capacity should seem large in proportion to her size, with a deep and wide barrel and a large, deep heart girth. There it is. If all this sounds a bit like mumbo jumbo, well, to some extent it is. These are important characteristics to observe; but for a family, a healthy, docile, good-producing but ugly animal might meet your needs better, and cheaper, than one who would be graded excellent on a dairy judge's scorecard but is a pain to handle. Overall, I'd say keep these conformation points in mind but don't place too much emphasis on them when making your final decision.

Now comes the big one. Look at, and especially feel, the mechanism you're going to be looking at and feeling every day, two times a day, three hundred days a year—her udder. The first misconception about udders is that the bigger the better; it ain't necessarily so. Ask what the cow's production level is, and if it seems about right for your needs, just keep on going.

Look at her udder from the rear and from the side. It should be well attached to her body (both in front and rear), not huge and floppy. Broken-down udders—even to the point where the cow could step on it or easily injure it on things—are not uncommon and are not good. If you go by the general guide that the floor of the udder should be above the hocks, you'll probably be safe. The udder should be symmetrical (avoid lopsided udders), wide and deep, and although you should see some cleavage between the halves (from the rear), there should be no cleavage between quarters (from the side).

Fig. 3-1. Parts of a Dairy Cow

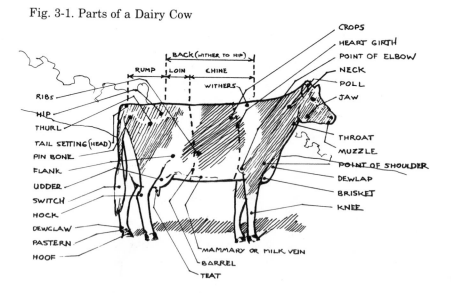

Now look at the teats; for easy hand milking they should be relatively long, squarely placed, wide apart, uniform in size, and free from cuts. There also should be four. This last statement isn't quite so dumb as it sounds; some cows have extras (supernumerary teats) which should have been removed when the cow was a calf because of the increased likelihood of disease.

Feel the entire udder; it should be soft and pliable, with no strange lumps or bulges, and seem equally full in all quarters. (If it's close to milking time, the udder may seem fairly hard.) The udder should be well veined; blood supplies all the precursors of milk, so a good blood supply is important for good production. If the cow is dry, it will be a little harder to judge her udder, but the basic judging principles remain.

Find out how old the cow is and how many calves she's had. A cow is not considered mature until she's six years old, and she doesn't reach her full production level until then. For a commercial herd, however, six is old, so a cow culled from a commercial herd for that reason might be fine for the homesteader. Cows have been known to live to ripe old ages of more than twenty, but they begin to decline in production at around nine. The cow should have started calving at around two or two and one-half years old, and then had one calf approximately every year thereafter. If there's a big gap, find out why (it could be a breeding or calving problem or she may have suckled a calf for a particularly long time and not been promptly rebred). Ask if her calving has been easy. See if any of her calves are around; are they thrifty and vigorous looking?

Is the cow bred? To what breed? If she *is* bred, find out when she was bred and figure out when she'd be due. Most cows freshen in either the spring or fall, every year, and it's hard to change them over. Does her calving season fit in with your schedule and needs? If she was bred artificially, there should be a breeding slip with the date and sire on it. If she was bred naturally, ask where you might see the bull. She should have been bred within ninety days of her last calving; if she wasn't, why not? Is the cow registered? This is not a big issue, but a registered animal is likely to be more expensive, although this expense can be offset by the fact that her calves may be worth more if you sell them. Ask if there are any records on the cow and her ancestors; many farmers belong to testing programs and will have official records of the cow's milk production, breeding, and calving.

Ask about diseases; there are many to be concerned with, but it's good to know whether the cow ever had mastitis and how badly. If she's a Jersey, ask about milk fever, a disease more common among Jerseys and usually repeated by the same cow at each calving. It might be worth your while, if you're pretty sure you want the animal, to have a vet check her over. A checkup will cost around ten dollars, but it will help reassure you that the cow is in good general health and will determine her stage of pregnancy or breed-ability if she's not bred.

What about her feed? How much has she been getting and what kind? This information is helpful in a few ways. First, it gives you a good idea of what to feed her if you haven't had any experience. Second, if you do buy her, you'll want to avoid switching her feed too quickly from what she had been getting to what you'll be feeding. Third, figure out if it's a reasonable amount of grain for her production level (see "Feeding" later in this chapter). If she's getting tons of grain, not producing much milk, and looks like a sack of bones, you know something's wrong—which reminds me of one final point: dairy animals are skinny looking and they should be. Don't buy a fat, sleek animal; she's the one who's going to have breeding problems, and she's the one who's "putting her feed on her back rather than in the milk pail." Certainly the cow should not be painfully thin, but part of a dairy character as opposed to a beef type is that long, lean look.

Lactation

The lactation cycle may seem a bit of an esoteric subject to the beginning stockkeeper, but if you think about it (a quarter of a second should do it), you'll see immediately how critical is an understanding of this pattern. The whole reason for keeping the cow is, after all, milk production, and the knowledge of just what is going on during that production will determine most of your management decisions.

Picking a point to begin describing a cycle is always tricky, but I doubt anyone will argue with "the lactation cycle begins at calving." At calving time, the cow's udder will be swollen with first milk, or colostrum. Colostrum is very diffferent stuff from regular milk and in very significant ways: it's very thick, more than four times richer in protein, two times higher in fat, and ten times higher in Vitamin A. All this nutrition helps the newborn calf get off to a good start, and tremendously aids in fighting illness. Much of the additional protein is in the form of antibodies, which can be absorbed through the milk by the calf during the first twenty-four hours of its life, thereby giving it a passive immunity against many diseases. Vitamin A, found in very small quantities in the newborn's reserves, is consumed and absorbed and then helps in the control of infectious diseases. For this reason, it is very important that the calf get colostrum as soon as possible after birth—within four hours is best. Colostrum is produced for a few days as regular milk secretion gradually increases, until after four or five days you're back to regular milking.

Milk yields increase rapidly after calving until peak production is reached approximately three to six weeks later. After the peak, a gradual decline in milk yield occurs. The rate of this decline is termed persistency, and a cow is considered persistent if her production does not drop too rapidly. If the persistence is satisfactory, each month's production after the peak will be 90 percent of that of the preceding month. (A typical lactation curve is shown in Fig 3-2.)

Generally, the lactation period is 305 days or 10 months, followed by a 2-month dry period; then the cow calves, and the cycle starts again. Three pieces have been left out of the picture, though: (1) For the cow to calve again, she must be bred—when, and does it affect the curve? (2) What ends the lactation cycle after 10

Fig. 3-2. Lactation Curve

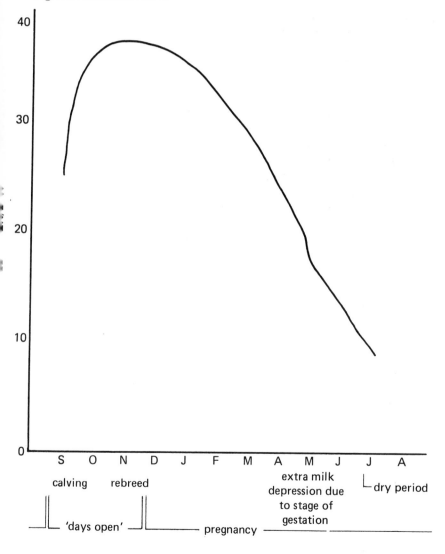

months? (3) What determines the height of the peak and the overall lactation yield?

First, ideally, the cow is bred 60 days after she calves, although more usual is 90 to 120 days. Sixty days open (unbred) allows for a 12-month calving interval—the goal. (Now you sharpies out there will have caught on to the fact that 9 months of gestation and 2 months open makes for an 11-month calving interval. This timing is possible, but difficult, as gestation is a little longer than 9 months, and cows don't always come into heat exactly on day 60 after calving.) Most commercial dairies are struggling to improve their 13- and 14-month calving intervals, so don't feel bad if you can't get your cow bred back right on time.

As you can see from the graph, during approximately eight months of the ten-month lactation, the cow is pregnant. Pregnancy has a depressing effect on milk yield, which is felt most in the ninth and tenth months of lactation, corresponding to the seventh and eighth month of pregnancy. (A friend of mine once titled a chart on this subject "The Depressing Effects of Pregnancy." Being pregnant at the time, I didn't think it was very funny.) The effect of this reduction is not too significant on the overall milk yield, however, as the cow is about to be dried off anyway—which brings us to the second point.

The dairy person ends the cycle at ten months by drying the cow off, making her stop milk production. A dry cow, then, is one who is not producing milk, usually because she's in the last two months of pregnancy (although it could be for other reasons). I'll never forget the time at the farmer's market when, during a particularly wet year, a woman asked why I had raised the price of some milk-containing item, and I explained, "Well, my cow is dry. . . ." She seemed very surprised and said, "What? With all this rain?" *That* is not one of the other reasons a cow may be dry.

Anyhow, it is agreed by all and sundry that a dry period of two months is necessary for maximum production in the succeeding lactation, although the reasons for this are less than crystalline. Some say it has to do with replenishing the body stores to meet the demands of the peak lactation soon to come. Others say it is related to the rebuilding of the udder's secretory tissues. All agree, however, that it should be done; *how* to do it can be debatable.

Most commercial dairy people use the cut-throat technique: completely stop milking the cow on the given day around sixty days before she's due. Cut out grain completely, and if possible, restrict water intake. The pressure in the udder from the unextracted milk will prevent further milk production; eventually, the milk remaining in the udder will be absorbed by the blood. This is considered the best technique for the healthy cow. Many cow owners, though, use other methods, primarily variations on the technique of gradual withdrawal. In this method, one milks every other milking for a while, then every third milking, and so on, gradually reducing the number of milkings until the cow is dry. The theory behind this technique is that even though eventually you're just going to stop milking, by that time the cow will be producing very little milk as a result of the pressure building up in the udder after each of those missed milkings. Simply milking the cow partially at each milking, another system used by some, is a poor method because the decrease in udder pressure at each milking allows more secretion to take place and just prolongs the process interminably.

Betsy, our Jersey, is very difficult to dry off. The gradual technique fails completely; no matter what we do, short of starving her completely I suppose (we never tried that), she just saves up the milk we didn't take out and generously delivers it all at the milking of our choice. The cut-throat technique almost works with her, and it is certainly faster. We cut out the grain and don't milk her at all. After a few days, though, when she's waddling around with a huge udder, I milk her out. With luck, after the one sympathy milking, if she fills up again, it's less, and then we just leave her alone.

Third, what affects the peak and the overall yield during lactation? Actually, the two are closely related, as the height of the peak determines the overall lactation production. It follows, then, that dairy people aim to have their cows peak at the highest possible level. In determining the peak and the consequent milk yield, there are both physiological factors (the genetic makeup of the cow, age, number of past lactations, pregnancy) and environmental factors (health and body condition, nutrition, ambient temperature, season, milking intervals, and dry period). Clearly, the dairy cow owner has control over only the environmental factors (except, of course, that physiological factors can be changed by getting another cow), which also contribute significantly to daily yield fluctuations.

Reaching the peak milk production possible for an individual cow (no cow can produce more than her genetic capability, no matter what you do) is achieved by manipulating the environmental factors to the optimal combination. Proper feeding, a major management tool, is covered later in this chapter. Health and body condition is another important factor; only a completely healthy animal will be able to fulfill her genetic potential. Cows must be in good body condition at calving (not thin) and, as mentioned earlier, have had a dry period in order to peak their highest. Overconditioning animals, particularly first-calf heifers, is a poor practice, as udder tissues will fill with fat, and cows will be more susceptible to various metabolic diseases after calving.

The season of freshening also affects milk yields. In the northern United States, fall and early winter fresheners produce more milk over their lactation than do spring fresheners. Much of this difference is related to feeding and management and may not be as applicable to the homesteader as to the commercial dairy farmer; however, there are two reasons for a fall calving that the small producer would do well to keep in mind. (1) It is best to avoid having the cow peak during the summer, as would a spring freshener, because the height of the peak will be depressed both by the higher summer temperatures and by the undernutrition that frequently exists at this time of year because of poor pasture conditions. (2) A significant boost in milk production occurs when the cow is turned out to fresh pasture in May and June. When this boost occurs toward the end of lactation, when milk yields are declining, the extra milk has a big effect on the overall lactation yield. This is really a boon for the homesteader. Just when you're thinking of cutting back on your cheesemaking, you turn the cow out, and woosh! you're back in business. One might argue that when you turn the cow out to fresh pasture when she's peaking, she'll achieve that peak with less concentrate feeding. This is true if you've got good pasture, but you're still battling with heat, insects, etc. It also means that the cow will be calving in the early spring, a time of year when other work is piling up, meaning less time and attention for the cow and calf. I also like fall calving for a very selfish reason. You love having your cow and you love having all the milk and goodies, but after ten months of daily milking, you're looking forward to a vacation. And me, I'd rather have my vacation in August, September, or October, the time the cow would be dry for a fall calving, than in February or March, the dry time for a spring season. Selfish and calculating, maybe, but I sure love those unpressured autumn evening horseback rides during "milking time."

Management

Housing

Any consideration of housing for your cow or cows clearly must begin with an appraisal of your resources. Do you have an old barn you are remodeling? Is it spacious or cramped? What other livestock do you have or plan to have? Are you going to build? What animals do you plan on housing together?

Most family cow owners have only one or two cows and one or two calves. Most, too, use an existing old barn or shed for housing their animals. But regardless of whether you have one cow or twenty, or are renovating or building, there are basic components to remember in housing dairy cows.

1. Comfort of the cow. In order for the cow to stay healthy and produce to her potential, she must be reasonably comfortable (magazines and color TV will do fine). Primarily, this requirement involves protection from the elements: wind, rain, snow, sleet, etc., and from temperature extremes if possible. Cold does not adversely affect milk production until temperatures drop below 5°F. The low temperature may cause a production drop, but as long as the cow is dry and out of the wind, it poses no health hazard. Comfort also means not forcing the animal to lie down in wet bedding. Keep bedding clean and dry. Ventilation without drafts should also be considered.

2. *Avoidance of injuries*, especially to teats and legs. Not only does this requirement involve keeping the area free of nails, wire, broken glass, etc., but it also means designing the area properly. Make sure stalls are big enough (if boxes) or wide enough (if stanchions) to avoid stiffness, lameness, sore or swollen hocks, or stepped-on teats. Make sure the area is kept dry enough to prevent slippery conditions.

3. *Reducing exposure* of the animal to stress and disease.

4. *Feeding system.* A feeding system should be set up so the cow can easily consume as much as she should.

5. *An environment* for the production of clean milk.

6. *Human comfort.* Some consideration can be given to the comfort of the humans working around the cow: things like ceiling height, dampness, temperature, slipperiness, etc. (but no magazines).

7. *High labor efficiency.* This is a factor often overlooked by small producers, who seem to forget that being a noncommercial enterprise does not preclude efficient labor use. A dairy farmer will quickly get indications of poor labor efficiency, namely, poor income. But people who tend to think of their labor as free, and so never plan for efficiency, may never learn that it's taking them three times longer than it should to produce that gallon of milk. Instead, they may just slowly become disillusioned, deciding that, after all, it's "too much work," and never know that the "too much" was due to poor planning. All the chores attendant on milking one cow—milking, pen cleaning, feeding, and watering—should not take more than an average of fifteen minutes per milking. Milking more than one cow should decrease the per cow time slightly, *if* you're efficiently organized.

8. *Economic feasibility.* Here's another item often overlooked by the overenthusiastic. If getting a barn built for all your animals will cost more than you spend on food for ten years and will bring in no income, you might want to reconsider your plans. Again, the dairy farmer finds out painfully when economic decisions are wrong, but the person going blithely along, never considering the econom-

ics of his or her situation, will never know and is, frankly, a fool. You should know what your enterprise is costing.

9. Manure disposal. Some system of manure disposal should be planned along with housing. Most homesteaders use a pitchfork and wheelbarrow for cleaning pens or stalls; the manure is then taken somewhere and dumped. Eventually, the pile, which has grown over time, is spread over fields or gardens, either by pitchfork and wagon or truck or by manure spreader. Some thought should be given to the location of the manure heap in terms of flies, odor, and runoff.

These are the general principles to be remembered when planning any sort of dairy housing; a few specific details, however, can be added.

If you have only a few cows, whether housed in stanchions or boxes, and you have a choice in floor materials, go with dirt rather than concrete or wood. All types of floors need bedding, but dirt needs the least, resulting in less work for mother. Wood absorbs urine and odors; dirt and muck get trapped below to provide lovely bug-breeding grounds; it's very difficult to clean. Concrete is cold on the body and hard on the feet and legs; a lot of bedding is needed because all the urine sits nicely on top of the concrete with nowhere to go but up. It has the advantage, however, of being easily cleaned and disinfected. Good old dirt, though, absorbs without odors. It's soft to stand on. Occasionally it will need replenishing—some dirt is always scraped out with the bedding—but it is by far the cheapest flooring you can have and needs the least labor.

Cows in box stalls or stanchions should have fresh, dry bedding provided as needed (after the wet bedding is removed). Bedding can be any clean, dry, and absorbent material, usually sawdust, shavings, or straw. Some people like the system of having a manure bed build up during the winter to provide heat for the animals; the pack is then cleaned out once or twice a year. This method is fine, as long as the top layer is dry, but to keep that top layer dry, large amounts of bedding are needed.

Deciding between a box stall or a stanchion as the basic housing unit for your cow or cows depends largely on space available. If there is room for box stalls (about eight by ten feet each) or a larger "loose housing" area for a few cows, I prefer this housing over stanchions, simply because I like to see animals able to move around, and they do seem to be happier if they can. Where space is limited, it may be necessary to use stanchions; the swinging type is considered more comfortable for the cow than the fixed type. If your cow is housed "loose," have an area where she can be tied up, if necessary, for vet visits or breeding; it can be the same place where you milk her.

Although cows can be hand milked in their stalls, it is preferable, for reasons of sanitation, to have a separate milking area convenient to the stalls. Feeding and watering systems should be such that the cow can easily eat what she's fed, and you can easily feed her. A triangular hayrack across one corner of the pen keeps the hay off the floor in a box stall. A shelf to catch the hay leaves is a nice added feature for a hayrack, since leaves, normally lost underfoot, are the most nutritious part of the hay. In stanchions, hay is frequently fed off the floor; grain is fed either while milking or in a bucket. Feeding grain off the floor is wasteful and encourages rodents. The water bucket, assuming you don't have an automatic watering system, should be wide enough for comfort and should hold at least five gallons. The feed bucket, too, should be large enough to prevent waste; narrow feeders encourage the animal to pull its head out while eating, spilling half your precious grain all over the place. All feeders and waterers should be accessible from outside the pen for easy labor. Remember that cows are grazers and naturally eat and drink with their heads down; don't place the feeding and drinking equipment too high.

Ventilation in the cows' area is important; cows give off a lot of moisture, which builds up in a poorly ventilated area causing damp, unhealthy conditions for man and beast. If direct natural sunlight can be provided for the cow, do so. Sun is great for sanitation and odor and is an important source of Vitamin D, especially during those long winter months when the cows are continually indoors.

Feeding

Feeding a dairy cow is much like feeding any other livestock; she must be fed, each day, a combination of feeds that will provide the necessary nutrients for her maintenance and production level. Of course, a good dose of common sense and judgment has to go along with calculated rations; no book can feed your cow. "The eye of the master fattens the stock"; that old saying can be rephrased, "The eye of the master milks the cow."

What you feed your cow will depend on (1) how big and what age she is: how much she needs just to maintain herself and grow, if necessary; and (2) what she is producing. For some part of the year, she is producing just milk; the question to answer is, "How much?" For a large part of the year, she is producing milk plus growing a calf; then milk production and stage of pregnancy determine her feeding.

All dairy rations, for all stages of production, are built around the roughages available, since they are still the cheapest source of nutrients. For the dairy farmer, it usually means dealing with hay, corn silage, or haylage. For many small producers, though, the only roughages to consider are hay and pasture.

A cow will eat approximately 2 to 2½ pounds of good to excellent hay per 100 pounds of body weight per day. For a 900-pound cow, this amount is 18 to 23 pounds of hay per day, or approximately half of a 50-pound bale. (To estimate the weight of your cow, see Table 3-1.) Over a year, that figure comes to 4½ tons of hay. However, presumably there is pasture available; when the pasture is good, no additional forage need be provided, so usually you figure your hay needs only for the winter feeding period. (We estimate 2½

Table 3-1. Estimating Weights of Dairy Cattle by Using Heart-Girth Measurements

1. Measure the animal directly behind the front legs.
2. Add the following number of inches to the measurement:

	JERSEY	GUERNSEY	AYRSHIRE	HOLSTEIN
Less than 3 yrs. old	0	2	2	6
3-4 yrs. old	2	4	4	8
5 yrs. old or more	2	5	5	9

3. Find the total in the table below and read off estimated weight:

TOTAL INCHES	WEIGHT	TOTAL INCHES	WEIGHT	TOTAL INCHES	WEIGHT
50	475	60	666	70	886
51	493	61	687	71	910
52	511	62	708	72	934
53	530	63	729	73	958
54	548	64	751	74	982
55	567	65	773	75	1,007
56	586	66	795	76	1,032
57	606	67	817	77	1,057
58	626	68	840	78	1,082
59	646	69	863	79	1,108

4. For heifers, use the original measurement:

INCHES	WEIGHT	INCHES	WEIGHT	INCHES	WEIGHT
25	52	30	87	40	196
26	58	31	95	41	211
27	64	32	104	42	226
28	71	33	114	43	241
29	78	34	124	44	257
		35	135	45	274
		36	146	46	292
		37	157	47	311
		38	170	48	330
		39	183	49	350

tons for our 7-month hay-feeding season for one Jersey cow.) Make sure you have enough space to store the hay you'll need during the worst part of the winter; looking for and moving hay in midwinter is a drag.

Judge the quality of the hay you're feeding. If it is extremely poor and you can't possibly get anything else, so be it; but it will be more expensive in the long run than providing good quality hay because of the extra supplementation the cow will need on poor roughage. With a very high-protein hay (i.e., excellent alfalfa or other legume), a low-protein concentrate is ample (10 to 12 percent) and can be supplied using corn alone. On the other hand, with poor grass hay, a 20 to 24 percent protein mix will be necessary. Good mixed hay should be supplemented with a 14 to 18 percent concentrate. Good hay, in addition to providing protein and energy, will provide adequate amounts of calcium and Vitamins A and D. Phosphorous in sufficient quantities, however, is not supplied in hay and therefore must be supplied by the concentrate (grains are high in phosphorous) or by a phosphorous supplement (a mineral mix).

Excellent roughage, liberally fed, along with water, salt, and phosphorous, will provide all the nutrients necessary for a cow to maintain herself and produce a small to moderate amount of milk (ten to fifteen pounds depending on the fat content). Even good hay, fed generously, might allow for some milk production. But for production above minimum levels or with roughages of the usual quality, some concentrate feeding is required. The big question is, how much?

The rules of thumb are always around to fall back on in a pinch: one pound of grain to three pounds of milk for Jerseys and Guernseys, and one pound to four pounds of milk for the other dairy breeds. However, these rules don't take into account the quantity or quality of the roughage being fed or the production level of the cow; they tend to overfeed low-producing cows and underfeed high producers. A more accurate determination of what to feed your cow can be made by using the grain-feeding guides (Table 3-2). These tables account for both quality and quantity of roughage, as well as for milk-fat production. They don't, however, tell you what protein level to feed. To figure that out, you can either use the rough estimates presented earlier, based on the roughage quality, or try to fit your situation into one of the descriptions in Table VII of *Morrison's Feeds and Feeding* (see Appendix I). Of course, the most accurate method is to calculate the protein necessary based on actual requirements, using the technique described in the Appendix. It takes only a little time to do the calculations, and it really is the best way to determine your cow's ration.

As you can see from the tables, the amount of grain required varies with the level of milk production. Don't make the mistake of basing the cow's feeding on her production at one point in the cycle.

Her milk yield changes with stage of lactation, and so should her ration.

Concentrate mixes can be put together in two ways: commercially, where you buy a premixed grain feed of a certain protein level; or mixing your own, either from purchased grains, home-grown grains, or a combination of the two. The advantages and disadvantages of the different systems are discussed in Chapter 1, as well as how to balance a ration when mixing your own.

Now, a note on using pasture as your forage instead of hay. Good pasture is the backbone of economical dairying, especially for

Table 3-2. Grain-Feeding Guides

Pounds of concentrate to be fed daily to cows:

I. Consuming 2½ hay equivalents per 100 lbs. body weight of excellent roughage

MILK FAT (PERCENT)	3.0	3.5	4.0	4.5	5.0	5.5	6.0
lbs. of milk daily							
10	—	—	—	—	—	—	—
15	—	—	—	1	1	1	2
20	1	1	1	2	2	3	3
25	2	2	2	3	3	4	4
30	3	3	3	4	4	5	6
35	3	5	6	8	10	11	12
40	6	7	8	10	12	14	16
45	8	10	11	13	15	17	18
50	10	12	13	15	18	20	21
55	15	17	18	21	26	28	29
60	17	19	20	25	29	31	33
65	18	20	22	26	31	ad lib	ad lib

II. Consuming 2 hay equivalents per 100 lbs. body weight of average quality roughage

MILK FAT (PERCENT)	3.0	3.5	4.0	4.5	5.0	5.5	6.0
lbs. of milk daily							
10	—	—	—	1	2	3	3
15	2	2	2	3	4	5	6
20	3	4	5	7	7	8	9
25	5	6	7	9	10	11	12
30	7	8	9	11	12	14	15
35	9	11	12	14	15	17	18
40	11	13	14	16	18	19	20
45	14	16	18	20	22	24	ad lib
50	17	19	21	23	26	ad lib	ad lib
55	21	23	25	27	30	ad lib	ad lib
60	25	27	29	34	ad lib	ad lib	ad lib
65	29	31	33	ad lib	ad lib	ad lib	ad lib

Table 3-2. Continued

61

Dairy Cows

III. Consuming 1½ hay equivalents per 100 lbs. body weight of fair- to poor-quality roughages

MILK FAT (PERCENT)	3.0	3.5	4.0	4.5	5.0	5.5	6.0
lbs. of milk daily							
10	3	4	4	5	6	7	8
15	5	5	6	7	8	9	10
20	8	9	10	10	11	13	14
25	10	11	12	13	14	15	16
30	12	13	14	16	17	18	19
35	15	16	17	18	20	21	22
40	17	18	19	20	22	23	24
45	20	22	23	24	26	27	28
50	23	25	26	27	29	30	31
55	26	28	30	31	33	34	36
60	29	31	33	35	37	ad lib	ad lib
65	33	35	37	ad lib	ad lib	ad lib	ad lib

the one-cow owner. Pasture is rich in vitamins and minerals, and at various stages of growth and with various plants, may be very high in protein. When there is an abundance of good pasture, cows can produce milk with very little concentrate feeding (Table 3-3 covers grain feeding on pasture); with good quality pasture, the protein level of the concentrate needn't be high (12 percent). Of course, protein levels in the grain will have to increase as pasture quality decreases. When thinking about feeding principles for a cow on pasture, substitute pasture quality for the hay or roughage quality mentioned previously.

Keeping your cows on pasture has lots of advantages other than providing a cheap source of feed. There is no stall cleaning, no bedding, the cow feeds herself (and happily) and spreads the manure at the same time. Cows on pasture are less prone to several metabolic disorders, respiratory problems, and nutritional deficiencies. They get plenty of exercise and soak up lots of Vitamin D, which they can store and draw on during those long months in the barn. The labor is less, the cost is less, and the animal is happier—what could be better? (Appendix II deals with pasture management.)

Whether the cow's on pasture or consuming harvested roughages, her body's needs for maintenance, lactation, and the early stages of pregnancy are met by using the preceding guides. During the last third of pregnancy, however, the fetus is growing rapidly and is beginning to make demands that must be met by the nutrient intake of the dam. Again, this amount can be calculated by using the feeding standards. Or one can be a little more casual about it and go by the rules of thumb for dry cows. (Cows are dry during the last two months of gestation—which is close enough to the last third.) If the cow is on excellent pasture at drying off time, and she is in good body condition, she will need no other feeding. If the

Table 3-3. Grain-Feeding Guides for Cows on Pasture

Pounds of concentrate to be fed daily to cows on:

I. Excellent quality pasture

MILK FAT (PERCENT)	3.0	3.5	4.0	4.5	5.0	5.5	6.0
lbs. of milk daily							
10	—	—	—	—	—	—	—
15	—	—	—	—	—	—	—
20	—	—	—	—	—	—	1
25	—	—	—	2	2	3	3
30	2	2	2	4	4	5	6
35	3	4	4	6	6	8	8
40	5	6	7	8	9	10	11
45	7	8	9	10	11	13	14
50	9	10	11	13	14	16	17
55	11	12	14	16	17	19	20
60	13	14	15	18	19	21	23
65	15	17	18	21	23	25	27

II. Average quality pasture

	3.0	3.5	4.0	4.5	5.0	5.5	6.0
10	—	—	—	—	—	—	—
15	—	—	—	—	1	2	2
20	—	—	2	3	3	4	5
25	3	4	4	6	6	7	8
30	5	5	6	7	8	10	11
35	7	8	9	10	11	13	14
40	8	10	11	13	14	15	16
45	10	12	13	15	16	18	19
50	13	14	15	18	19	21	23
55	14	16	17	20	21	23	ad lib
60	16	18	20	22	ad lib	ad lib	ad lib
65	18	20	21	ad lib	ad lib	ad lib	ad lib

III. Poor quality pasture

	3.0	3.5	4.0	4.5	5.0	5.5	6.0
10	—	—	1	2	3	4	4
15	2	3	3	4	4	5	6
20	4	5	6	7	8	9	10
25	6	6	7	9	10	11	12
30	8	9	10	12	13	14	15
35	10	12	13	15	16	18	19
40	12	13	15	17	18	20	22
45	14	16	17	20	21	23	ad lib
50	16	17	19	21	ad lib	ad lib	ad lib
55	18	20	21	ad lib	ad lib	ad lib	ad lib
65							

pasture is just good or her body condition is only fair, or if she is being fed good hay, some supplementation may be in order: two pounds or so of a 12 to 14 percent concentrate mix. If, however, the cow is thin, concentrate feeding should be increased to five or six

pounds per day. Not only will this feeding level take care of the
developing calf, but it also gets the cow in condition to meet the
demands of the upcoming lactation. You can see, though, that the
"eye of the master" plays a large role in feeding the dry cow. Even
if you calculate the ration from the standards, use common sense;
pad the ration if the cow seems thin, and pare it a bit if she seems
heavy.

After calving, cut the cow's grain for one day. Then, for the
first feeding, give her just one pound of concentrate (she can have
her hay). Thereafter, increase the grain allowance daily, but gradually (about one pound per day), until the cow is on full feed.

Other than roughages and concentrates, the only additional
feeds to be considered are water and salt. Plenty of fresh water
should be provided at all times; a cow may drink 100 to 300 pounds
of water a day—make sure she has it available. Salt, along with
other minerals, is frequently mixed into commercial concentrates,
or it can be added in the form of salt and mineral mixes. But by far
the easiest thing to do, and a good extra guarantee, is to have a
salt block available to the cow at all times. Do it. So many family
cow (or goat, horse, etc.) owners forget to provide salt; yet salt is as
necessary a nutrient as any of the others.

Feedings should be on a regular basis and always in the same
order. Generally, grain is fed at milking and hay afterwards, primarily because of the dust clouds produced by feeding hay before
milking, but also to avoid any off-flavors in the milk as a result of
the hay eaten. Some people argue against feeding grain while milking, claiming that the cow will be more likely to defecate at that
time. Others haven't found this to be a problem, or they feel that
having the cow eager to come into the milking area and standing
quietly while being milked is worth the occasional extra trouble.
You should realize, though, that if you feed grain right after milking, you might be dealing with a restless cow every time. Since
feeding takes up a good chunk of the time and care in dealing with
a dairy cow, you should make the entire feeding process as efficient
and wasteless as possible.

The economics of your rations should always be kept in mind;
you should know how much the feed is costing per gallon of milk
produced. It's very easy to be penny wise and pound foolish with a
backyard cow. Remember that the cow's maintenance requirement
stays the same whether she's producing two gallons or four. Spreading that basic maintenance cost over the largest amount of milk
possible will result in the most economical production. Economic
calculations should also be used to decide whether it's worth your
while to feed more grain to get more milk. If, for example, on three
pounds of grain (plus hay) the cow produces two gallons, but on
eleven pounds will produce four gallons, you're adding eight pounds
of grain to get two gallons of milk. If grain costs seven cents per

An excellent nine-cow, "million-pound" Ayrshire herd making maximum use of pasture. Owned by H. and L. Lahman, Elkton, Va.

pound, that's an added investment of fifty-six cents for a return of two gallons of milk. Whether that return is worth it to you is a personal decision but one you should be aware you may be making if you're skimping on grain. (Of course, the cow cannot produce past her genetic ability, no matter how much you feed her.)

Breeding

The time has come to breed your cow. Maybe you bought her open (nonpregnant), or perhaps she has recently calved. That little calf may be only two months old and still on mama's milk, but it's already time to think ahead to next year's crop. Before picking a sire, though, figure out *why* you're breeding the cow. Presumably the main reason is the obvious one: to keep the cow milking. But as long as you're going to get a by-product from your milking machine, it might as well be the most useful by-product possible. Are you interested in raising a replacement for your cow? Perhaps your cow is old, but has been terrific, so you'd like to keep a daughter of hers to milk in the future. Or perhaps your cow comes from very good stock and you'd like to sell her calf rather than raise it for yourself. Maybe neither of these situations applies, and instead, you think you'd like to raise some beef or veal.

If you're going for a replacement, remember two things: (1) you won't be milking her for 2 to 2½ years, and (2) it has to be a she. If you are determined on a purebred replacement or you want to sell a registered calf, be sure to choose the best bull you can—it pays. If you plan on beef or veal, you can try crossbreeding your cow to a beef breed sire.

There are a couple of twists here, though, depending on the breed of cow you've got. If you have one of the large dairy breeds, I would breed it to a good purebred bull of the same breed. Then, if you have a heifer, you've got a good replacement or sale calf; if it's a bull, it can still be vealed or beefed quite satisfactorily. On the other hand, if you've got a Jersey or Guernsey, I would almost always cross it, either to a beef breed or to a dual-purpose breed. Then, if it's a heifer, you can beef it or milk it, but more importantly, if it's a bull, it will make good beef. Purebred Jersey or Guernsey bull calves are just so small and grow so slowly that they simply do not make good beef animals, and the chances are quite good, after all, that you will have a bull. There is virtually no market at all for a purebred bull of these breeds unless it is of superior breeding, whereas a crossbred beef animal is more salable and more practical to raise.

My own feeling is that since it is relatively easy to find milk cows, I wouldn't take the chance on trying for an all-Jersey heifer. A good dual-purpose answer would be breeding to a Simmental, giving you a good beefy calf or a good milker; many Simmental cows milk just as well as Jerseys or Guernseys, and they're reasonably high in fat. Another good choice might be an Angus; they are good milkers (for a beef breed) and produce excellent beef. Since Jerseys are notoriously easy calvers, you could consider breeding one to a slightly larger beef breed; a Jersey can more easily deliver a big calf than her size would indicate. Last year Betsy had an Angus calf that was a good ninety pounds—quite a size for a tiny Jersey. That Angus, by the way, was such a beautiful heifer and grew so well on Jersey milk that we sold her as a yearling for a good price to an Angus breeder to use as a brood cow.

Once you've selected the breed of sire you must decide between natural service or artificial insemination (A.I.). The big advantage to natural service is that you don't have to detect the cow in heat. The big disadvantage is that you need a bull. Certainly, for one or two cows, you're not keeping a bull, but perhaps a neighbor has one he rents out, in which case you bring him to your place and let him do his job. But for the family cow owner, this can be a huge, costly hassle, easily avoided by using A.I. All you have to do is catch the cow in heat, call the technician, and he'll come breed the cow.

Sounds simple, doesn't it? Well, the "catch the cow" is the catch of the whole thing. It's not that hard, but it does take some vigilance and being in tune with the cow's behavior patterns. For a period of time we had only Betsy, and the first time around we were being very diligent about detecting her in heat. We were overscrupulous—she let us know loud and clear. One day my husband, Jerry, was bending over cleaning out the water tub. Betsy couldn't resist and promptly jumped on Jerry's back. There was no doubt about it, the cow was in heat. Fortunately, there are much less dangerous methods than this to detect the heat period (estrus).

A cow comes into heat approximately every twenty-one days, with estrus lasting about eighteen hours (the time she is ready to be bred). When the cow is first coming into heat, she may bellow, sniff at other cows, and try to ride other cows, but won't allow cows to mount her. Her vulva will seem moist and swollen and deeper colored than usual. There may or may not be a clear mucous discharge. Standing heat is characterized by just what it says: the cow will stand for the bull or to be ridden by other cows. She may bellow, seem nervous or excitable, and will ride other cows.

Well, this ride-other-cows stuff is all very well if there are other cows to ride, but what if you have only one cow (and don't want to volunteer your own back)? One thing to watch for is a drop in milk production; many cows will have slightly lower yields while in estrus. Look for mucous discharges that persist (frequently cows will have a discharge when they first stand up, but it will disappear shortly). Watch for a change in her behavior. But best of all, know *when* to watch for these signs. Most cows will come into heat around fourteen days after calving. (If the cow has not appeared to come into heat by the sixtieth day postpartum, have her checked by a vet.) Watch her around then. If you catch that heat, you're set. Healthy cows are very regular, and will come into heat every twenty-one days or so after that first one, so you'll know just when to watch for the signs. If you think you've missed a couple of heats, carefully observe the cow at each milking. At some point you may see a bloody discharge or blood on her tail. This usually occurs two to three days *after* the cow was in heat, but you can count from there when her next heat will be. It is desirable to breed the cow during her first heat approximately sixty days postpartum, allowing time for complete involution of the uterus, while maintaining a twelve-month calving interval.

The best time for the cow to be bred is toward the end of estrus. As soon as you see the cow in heat, call the A.I. technician. If the cow was seen in heat in the morning, he'll come in the afternoon. If it was in the afternoon, he'll come the following morning. The insemination technique is roughly this: You've picked the sire from those he has available. He takes a vial of frozen semen, thaws it, and puts it in a plastic rod. With his left arm (if he's a righty) he reaches into the rectum, locates the cervix (you can feel it below, through the wall of the rectum), and holds it in place. Holding the catheter in his right hand, he inserts it into the vagina, then guides it through the cervix, and deposits the semen just inside the uterus. Done.

Breeding work is not over yet, though. Watch your cow around the next time she'd be due back in heat. If she shows no signs of heat, fine, she's probably bred (although I always go on watching nervously, month after month, until she shows). If you want to be really sure that the cow is bred, you can have a vet check her forty days after breeding. If she comes back into heat, it probably means the breeding didn't take, and she'll have to be bred again. Approximately 5 percent of pregnant cows do come into heat again, though, so this time, just in case, the inseminator will not insert the catheter through the cervix, thereby avoiding breakage of the cervical plug which forms during pregnancy.

If the second breeding also doesn't take, have the cow checked by a vet. She may have any number of reproductive problems, including cystic ovaries, infections, etc., or the problem may simply be one of heat detection and optimal breeding time. If breeding problems persist, however, it may be wisest to replace the cow rather than try to fix her.

You started out the whole breeding process by deciding on the breed of the sire, but how do you choose among the sires of that breed? Usually the A.I. technician will have with him a pamphlet or brochure with information about the sires available. Sometimes there are pictures in the brochures, so you could go just by looks ("Daisy would just *love* Abe; she likes to see black on a muzzle"). But there is other information as well. The quality of the bulls has been measured and tested by using records of the bulls' daughters in various herds. How well the daughter produced compared with her herdmates' average gives information about her genetic ability to produce milk. In other words, here is a cow under the same management as all the other cows, but she is producing more milk— why? It can largely be her inherited ability to do so.

Here is a sample of some of the information you might see on a proven sire:

Milkboy Happy Abe
Superiority: +350M   +.3%   +15F
Confidence range: ±200M   117 daus.

What this record means is that Abe's daughters produced 350 pounds more milk during their lactation than did the average cow; the milk had a .3 percent higher butterfat test (if the average was 4 percent, these were 4.3 percent) and produced a total of 15 pounds more fat over the year. Of course, any of these numbers could be minuses instead of pluses. The number of daughters whose records have been used in proving the bull is given (117) and affects the confidence range; the more records, the more accurate is the estimate of the superiority. In this case, a confidence range of ±200M means that milk production superiority of daughters of this bull would fall between +150M and +550M. Clearly, the smaller the confidence range, the surer the bet.

Look at the item you want to improve—do you want to select for milk production, total pounds of fat, or butterfat?—and choose the sire accordingly. Some sire reports will also give information about udder, legs and feet, milking speed, and disposition. Remember in weighing all the qualities that there are always tradeoffs.

Sometimes there will be young sires, not yet proven, who will have information presented only about their ancestors. These records read similarly, for example:

Lotus Noble
Sire: Bruce Westfield Cherry Noble—VG
    Superiority +650M +.3% +60F 263 daus.
    183 breed class. daus. av. 83.5

This is the name of the sire, who was classified officially by judges as Very Good. The last bit means that 183 of his daughters were judged on conformation as well, and rated 83.5 points out of 100, or Good Plus (not as good as Very Good).

Dam: Sup. over herdmates +6506M −.4% +243F
    3 recs. av. (2X 305d ME) 17,818M 4.5% 795F

You should know now what the first line means. The second line means that the bull's mother had three records: she milked an average of 17,818 pounds of milk at 4.5% butterfat and 795 pounds of fat. The records were based on milking twice a day for a lactation of 305 days, with standardization to a Mature Equivalent basis. The age of the cow affects her milk production, so all records are standardized to what the cow's production would be if she were a mature animal.

I always find it fun to go through the sire charts (especially the pictures) and pick out characteristics I'd like to see in the next generation. But the key to it all is to pick the best bull possible, then take care of mama properly so she can produce a healthy calf.

Calving

Calving is one of the most exciting events on the small farm. There's your cow—big and huge and dry. After two months of not milking, you're dying to get your wrists back in action. And of course, you're anxious about how things will go and what the calf will be, heifer or bull. Usually things go along smoothly as mother nature planned, but sometimes there is trouble, and occasionally even tragedy. In any case, it's best to know what to expect and what could go wrong.

Gestation length in the cow is about 280 days, with a few days leeway on either side. It varies slightly among breeds: Holstein 276, Ayrshire 278, Brown Swiss 290, Guernsey 284, and Jersey 279. Most cows deliver close to their due date, so if you know exactly when the cow was bred, you'll have a pretty good guess at her calving date. The signs of approaching calving can be textbook clear, or they can slip right by without your noticing a thing.

A few days before calving (although sometimes not until right after), the cow will bag up; that is, her udder will fill with colostrum. (If the udder seems unduly congested several days before the calving date, try rubbing the udder twice a day with balm.) As calving time approaches, the cow may refuse feed, seek a quiet spot, and have a mucous discharge. In addition, she may have a tendency to keep looking back at her flank, and her tail head will seem to sink down into her back as the ligaments loosen up in preparation for calving.

There are three stages to a normal calving. Stage 1, the preparatory stage, lasts two to six hours. The calf rotates to an upright position (its spine parallel to the cow's), uterine contractions begin, and the water sac is expelled. Once the water sac is expelled, delivery (stage 2) should take place within an hour. The cow is usually lying down but may be standing. The calf enters the birth canal, the front feet and nose protrude, and quickly and quietly, the calf dives out into the world. Stage 3, cleansing of the cow, is the expulsion of the placenta and occurs within six to eight hours.

If all goes well, the cow will start licking the calf immediately, within twenty minutes to a half hour the calf will stand, and within an hour, nurse. If mucous seems to prevent normal breathing, it should be wiped from the calf's nose, or if the calf seems particularly clogged up, it can be picked up by the hind legs and shaken. If by some chance the cow does not lick the calf, dry it off by rubbing it briskly with a towel. Be sure to dip the calf's navel thoroughly in an iodine solution before the navel dries out. The navel cord acts like a wick for bacteria, so the dipping is an extremely important step in the prevention of disease, especially one called navel ill. Generally the cow and calf should be left alone, but keep an eye on them to make sure the calf nurses within two hours. If it hasn't, you'll have to help the poor calf figure out what it's all about.

Calving should take place in a clean, dry environment, and one of the cleanest environments going is clean pasture. As long as the weather isn't wet and hideous—even if it's cold—it's perfectly okay for the cow to calve outside, and it's certainly preferable to an inside pen that's drafty and dirty. Of course, the big disadvantage to having the cow calve outdoors is that she's harder to keep track of and care for in the event of difficulties. Keeping her in a small outside paddock, though, can help offset that problem. If the cow is to calve indoors, move her to a large (ten by fifteen feet), cleansed, disinfected, and freshly bedded box stall two or three days before she's due. She can have feed and water as usual up to calving. After calving, the cow's grain should be withheld for the first day, although she can have all the hay she wants.

The problems which can, and do, crop up occasionally during calving are usually associated with abnormal presentations of the calf. Most of these problems require experienced help; either a vet or an experienced dairy person should be called. Unless *you're* experienced, don't try to solve these problems yourself unless you absolutely must (except in the case of a breech presentation). If you must go into the cow, wash your arm well with soap and water, and either put more soap on it or some other lubricant. Correcting most abnormal presentations involves pushing the calf back slightly and rearranging limbs, but some are much more complex. You may have read *All Things Bright and Beautiful* at least ten times, but get experienced help if possible.

Anterior presentations take many forms (Fig. 3-3). One or both forelegs may be bent at the knee, in which case the trick is to wrap your hand around the knee and try to unbend it. Or a foreleg may be crossed over the neck. The head may be out of position: instead of nose first, the head may be bent back and up. A calf may present posteriorly: on its back and rear end first. Or it can be in an upright position, but folded and backwards: a buttock and hock presentation. A breech presentation is the easiest to assist and should be assisted right away. The calf is presented hind legs first, but it is all stretched out, so there is no unfolding to do. (You can tell it's the hind legs because the hooves will be pointed down.) The aim is to get the calf out as quickly as possible; because of the calf's position, as soon as the cow starts to push the calf out, the navel cord is compressed, cutting off the oxygen supply. Grasp the calf's legs, and with each contraction, pull; don't try to pull against a contraction. When the calf comes out, turn it upside down and shake it to clear its lungs and stimulate it, and rub it vigorously with a burlap sack, making sure its nose is clear.

How do you know when the cow is having trouble? Keep an eye on her; if she has been straining for an hour with nothing to show for it, get on the phone fast. Never hestitate to call the vet and describe the situation.

One additional calving problem may involve the cow's cleansing. Usually the placenta is expelled within a few hours; however, frequently the cow eats it, so you never actually see it lying there. If you *know* she hasn't cleansed within seventy-two hours, call the vet. If she has had ample opportunity to gobble up the placenta while you weren't looking, you can be pretty sure she's cleansed if everything else went normally, and there are no unsightly things hanging out, and no sickly looking discharge.

If something happens and the calf dies, the cow should be milked out immediately, then put on the regular milking routine. Betsy lost one of her calves once, but her mothering instinct was so strong that we couldn't milk her out; she knew to let her milk down for a calf, and only a calf, and wasn't going to be fooled by a wet rag. The next day her udder was huge and hard, but she still wouldn't let her milk down. Finally, we did the only thing we could: we got her a foster calf and the problem was solved. (Fortunately,

NORMAL PRESENTATION

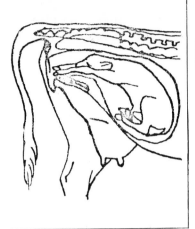

ABNORMAL PRESENTATIONS

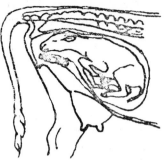

Anterior presentation;
both forelegs bent at
knees

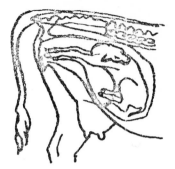

Anterior presentation;
head upward and
backward

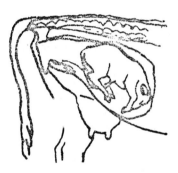

Buttock and hock
presentation

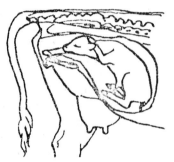

Anterior presentation;
foreleg crossed over
neck

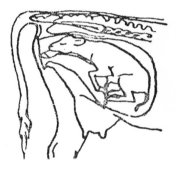

Anterior presentation;
foreleg bent at knee

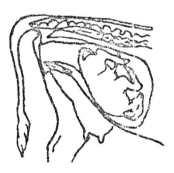

Posterior presentation;
fetus on back

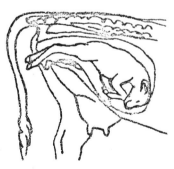

Breech presentation;
hind legs first

Fig. 3-3. Abnormal Presentations of a Calf

the calf was a 110-pound Holstein, happy and able to handle the bounty Betsy showered upon him.)

After the calf has nursed its colostrum, it's time to decide how long you'll leave the calf with its mother and what you'll be feeding it.

Raising Calves

Now that your cow has calved, there's a decision to be made (after deciding whether you're the proud parent of a heifer or a bull). What are you going to do with it? If you bred your cow to produce beef for the freezer, that's one thing. If you were hoping to raise a dairy replacement, but got a bull, that's another. (If you got a heifer, congratulations.)

Right off the bat, if you don't want the calf you can sell it as "bob veal" or use it yourself for veal. Or if you're not too keen on raising the calf to full-grown beef size but can keep it around for a while, you could raise it as milk-fed veal (200 to 300 pounds). However, for both the replacement heifer or for the beefer, you'll have to make long-range plans.

In all cases see that the calf has its colostrum within two to four hours of birth, and make sure mama is doing well. The calf can then nurse the cow for almost any length of time, from zero to six months. You *can* take the calf away from the cow immediately at birth, milk the colostrum from the mother, and then feed it to the calf in a bucket. Some say that this method lessens separation problems later and makes it easier to teach the calf to drink from a pail. I'd rather be sure the calf got off to a good solid start, though, with much less of my labor. In other words, leave the calf on the cow until the colostrum is gone (two to four days)—no buckets to worry about cleaning, no schedules to keep, nothing. After three or four days, you can separate the cow and calf as far apart as possible, out of each other's sight, and preferably, hearing as well. (Out of each other's hearing may be difficult. Out of your own hearing, even more desirable, is probably impossible. We put calves in one barn and cows in another, three-hundred feet apart; we had stereo mooing for two days.)

Put the calf in a clean, dry, well-bedded, draftless pen, ideally with natural light. The calf can now be fed out of a regular bucket or out of a nurse pail (a bucket with a nipple), either of which must be kept clean. I like the nipple pails because they're easier to teach the calf to use, and they gratify the sucking instinct. With a little warm milk in the pail, gently guide the nipple into the calf's mouth; usually the calf catches on immediately (you might have to squeeze a few squirts into its mouth). Then hang the pail from the wall or gate at around the calf's nose level for each feeding and remove it when the calf is finished (otherwise there may be a problem of sucking in air). To use an open pail, put some warm milk in it, and then get the calf to suck on your fingers (it's usually quite easy; just slide them right in there). Just before the calf sucks all the blood out from your fingertips, gradually lower your hand into the milk, the idea being that the calf will suck up some milk along with finger. You might have to try this a few times; just be patient. Remember that the calf naturally looks for its meal somewhere above nose level, and that it instinctively butts and bumps the milk source. If you have trouble, though, hunger is a good teacher; try again after the calf hasn't eaten for eight or ten hours.

At this point you can be feeding either mother's milk or milk replacer; a lot depends on how much natural milk you have available. If you have a surplus, mom's milk has several advantages over replacer: (1) just like with people, it was designed for the individual calf; (2) if you feed the calf at milking times, there's no milk warming to do; (3) you don't have to mix up solutions; (4) some people claim it produces superior veal. Milk replacer is advantageous if (1) you need all the milk the cow is producing, or (2) the cow's milk is too rich for the calf. It is said that too rich milk will cause scours in a calf, and it is probably true. But with 6 percent Jerseys and several calves, I've never seen it. We even raised a Holstein on 6 percent milk fat. Not only did he not scour—and surely his digestive system wasn't used to that richness—he grew like a weed. (One Holstein calf, however, does not a statistic make.) With careful management, you should be able to avoid having your calf scour, but switching to milk replacer (or a milk lower in fat) might help should it be a problem.

In either case, the first few days feed five to six pounds of warm milk daily, spread over three feedings if possible. Calves can learn to drink cold milk from the start, but it can result in more digestive upsets. More important than warm or cold, though, is to keep the temperature of the milk the same from feeding to feeding; if you start with cold, stick to it. If you're feeding warm, don't be lazy one day and suddenly switch to cold. For an extra large and vigorous calf, the milk level can be upped to eight pounds per day. Be aware, though, that there is more danger from overfeeding than from underfeeding. As the calf grows, increase its milk allowance gradually—about one pound of milk daily per ten pounds of liveweight is the rule of thumb. Weak calves should actually be fed less (and over more feedings), and the best of calves not more than twelve pounds per day. Should scouring start, immediately cut this amount down to half until the calf is recovered (see "Diseases" later in this chapter).

The milk-feeding regimen should be kept up for at least two months. At this point, the different reasons for raising the calf will result in different feeding programs. If the calf is being raised for

If barn space is limited, this calf hut can make excellent housing,
even where winters are severe. Made of four sheets of plywood, it is
simply a box, 4 feet by 4 feet by 8 feet, with a dirt floor and a
fenced yard. Half a panel is placed halfway down the box to create
a more sheltered inner space. The hut must be kept well bedded
and dry, with the opening oriented away from the prevailing winds;
hay stacked against the sides will help insulate it.

veal, it should get milk, and only milk, until it reaches 200 to 300
pounds (it takes about 10 pounds of milk for a pound of gain). When
veal calves are raised commercially, great care is taken to insure
that the meat stays white: iron is filtered out of all the water (the
calves are kept anemic), the calves get no exercise, and nothing but
milk replacer is fed. For the small producer who doesn't care
whether the veal is absolutely white, the calf can be allowed exercise (and unfiltered water). If you want milk-fed veal though, feed
only milk. If you're happy with "very baby beef" and you're short
of milk, you can feed some hay and grain.

The heifer replacement or the beef calf should have good,
green, leafy hay put in front of it (in a rack) starting at one week.
Use just a little hay; they'll only nibble at it, and it must be fresh
each day. (Calves don't like picked over hay; feed leftover calf hay
to other stock.) By about two or three weeks, they'll be doing some
real nibbling. However, it should be pointed out that your little
ruminant friend is not yet really a ruminant; it is as simple-stom-

ached as you and I, and must be fed accordingly. The rumen of the calf is underdeveloped and won't fully develop until the calf is about three months old. Consequently, the young calf is not equipped to handle roughages in any great quantity, and unlike the mature ruminant, both protein quantity and quality must be considered in formulating the ration (see Chapter 1). So, although the calf is eating hay, milk feeding should continue; the milk, in addition to legume hay, will make a ration adequate in protein and vitamins. At this point the calf becomes used to eating the hay, which stimulates development of the rumen.

At the same time hay is being introduced, the calf can be taught to eat small amounts of concentrates. If the calf is getting milk, the concentrate mixture needn't be high in protein. To teach the calf to eat grain, you can put some in the bottom of the pail after the calf has finished drinking, or rub some on its muzzle, or try to get the calf to suck some off your fingers and eventually lick it out of your hand. Once the calf has learned what the stuff is (and they usually like it), concentrates can be fed in a bucket or box. Until the calf is two to three months old, it can be fed all the grain it will eat. Over this age, four to five pounds per day, fed in two feedings, should be adequate. If the calf is still getting milk, the protein content can be about 12 percent. However, if the calf at two or three months is eating concentrates and hay well and in sufficient quantities, it can be weaned from milk completely and fed instead a high protein (18 percent) calf starter or grower mix. Be sure, though, that the calf is eating grain well before weaning it from milk. Water should be available (and actually should be provided from the time the calf is a month old).

There are several alternatives to this feeding plan. One, not a very radical change, is that you can feed skim milk instead of whole milk starting at about two to four weeks, gradually changing the ration from whole to skim. Once the calf is on all skim milk, it can be fed somewhat more than if it were on whole milk; up to eighteen pounds per day up to six weeks old, then as much as it wants after that. If a limited amount of skim milk is available, ten pounds of skim milk along with ample hay and grain will be adequate. The advantage of this plan is obvious: milk the cow; you get the cream; the calf gets the rest.

Another, more dramatically different, method is to leave the calf with the cow for longer (but varying) lengths of time. Some people have success with both milking the cow and leaving the calf to nurse, leaving them together all the time. The theory is that the calf is not drinking all that the cow is producing, so the calf gets its share, and you get the rest. Another technique is to keep the cow and calf separated, but only at night. In the morning, the cow is milked partially; then the two are left together all day—during which time, of course, the calf nurses. They are separated again at night. As the calf eats more and more hay and grain, the cow can be milked more and more completely each morning. (The opposite can also be done: keep them separated during the day; you milk in the evening.) A third method is to keep the cow and calf separated at all times, but put them together for a while at milking time. You milk the cow partially, and then put the calf with the cow for an hour or so to get the rest.

All these techniques are good if you can get them to work (no buckets to clean, no milk to mix or warm, etc.). We tried methods one and two, and found the second one more satisfactory. With the first, poor old Bess got too confused and didn't know who to let her milk down for—us or the calf. (She wouldn't consider both, which is what's needed for this method to work.) It didn't work for us, but I do know people who've been quite pleased with this system. The second method worked fairly well. We penned the cow and calf next to each other with a slatted wall between them (after the calf had nursed freely for a week); there was little trauma those initial nights of separation since they could still see, sniff, and with some effort, lick each other. The cow adjusted fairly well to being hand milked in the morning and then nursed by the calf; but again, she was a little confused and tended to favor the calf's stimulus over ours, thereby giving him a bit more than his share. When we finally weaned him completely (at three months) we just kept them in their neighboring pens all day as well as all night. There were some plaintive cries for a day or two, but it was relatively easy and painless for all.

(If you're raising a beefer instead of a heifer, feeding a beef calf is covered in Chapter 6; now we'll concentrate on heifer replacements.)

After six months, calves are a cinch to raise, as long as a few basic nutritional points are kept in mind. If fed plenty of legume hay, or if on good pasture, heifers will do well with a 12 to 14 percent fitting ration for a concentrate mix. With no legume roughage, the concentrate should be 16 to 18 percent. The amount of grain depends on the quantity and quality of roughage. With plenty of good roughage, feed two to four pounds of concentrates per day; with only fair quality, four to six pounds. Hay should be fed at a rate of eight to fifteen pounds per day. If you're lucky enough to have the heifers on excellent pasture, no grain need be fed. In the spring, when calves are first put out to pasture (they can utilize pasture starting at around three months), get them used to it gradually by continuing to feed some hay while they're away from the pasture (they might not eat the hay if you try to feed it to them while they're surrounded by lush green grass). Of course, water, shade, salt, and some shelter should be provided.

After heifers are one year old, they can be maintained up to three or four months before calving, either on lots of excellent legume hay alone or just ordinary hay plus two to four pounds of concentrates per day. Judge the condition of the heifers; they should be in good flesh, but not fat, for calving. Heifers can be housed satisfactorily in a three-sided shed if it's dry and out of the wind.

Size, rather than age, determines the time to breed your heifer. Holsteins and Brown Swiss are bred at about 725 pounds, Ayrshires at 600, Jerseys 500, and Guernseys 550. The expected ages at calving are 24 months for Holsteins and Swiss, 23 to 27 months for Guernseys and Ayrshires, and 22 to 27 months of Jerseys.

Calf Management

Two items should be mentioned: (1) vaccination and (2) dehorning and castration. If the calf is to be vaccinated against brucellosis, it should be done before it is six months old. Consult your vet on the need for this or other vaccinations, as local conditions determine what should be administered. Dehorning, if done at all, should be done by the time the calf is two weeks old. If you're raising a calf for veal, dehorning is unnecessary, and if the calf is being raised for beef, the need is debatable. The main reason for dehorning is to keep cows from damaging each other. If, however, you have only a few animals housed in roomy quarters and not squabbling over food, then horns are less of a problem, especially if the animal is only going to be around for a year or so (as in the case of a beefer). If you're raising a replacement heifer, though, be aware that if and when it comes time to sell her, horns will be discriminated against. One note of warning: young calves playing "butt" with you, especially at feeding time, are incredibly cute, and it's hard to resist playing butt back. However, an eight-hundred pound calf with horns playing butt is a whole different story. In other words, be aware of how you're treating a horned animal; it's the pets that often do damage to their owners.

If you choose to dehorn, either a caustic stick or an electric dehorner may be used. When using the stick, the hair over the horn bud is clipped and the button is rubbed vigorously, in a circular motion, with the caustic potash until it begins to bleed slightly. To make sure the stuff doesn't run into the calf's eyes or burn the skin, encircle the clipped area with Vaseline or lard.

The electric dehorner is a hot iron which fits over the horn button and destroys it by killing the nerves, thereby preventing further growth. A local anesthetic may be used around the horns, then the iron applied when red hot for a number of seconds until a deep, copper-colored ring appears around the horn. (The amount of time required for this procedure varies and requires experience with the particular iron.) The iron must be kept on for the proper amount of time (even though the calf may be screaming); an incomplete job results in scurs—partial horns. The horn button sloughs off in four to six weeks. The cost of an electric dehorner, unless you're doing several calves, isn't really worth it for the one-cow owner. A cheap alternative is to have the vet dehorn your calf when he or she is out at your place for some other reason; I wouldn't call the vet especially just to dehorn one calf.

Castration of bull calves used for veal is unnecessary; castration of bull calves used for beef is quite necessary and is covered in Chapter 6.

Calf Diseases

There are three major calf diseases to be aware of. One is *acute calf septicemia,* called white scours; it is characterized by horrible smelling, greyish-white, runny feces. The diarrhea weakens and dehydrates the calf, and death is fairly common. White scours is caused by a virus, followed by an *Escherichia coli* infection of the intestinal tract. Preventative measures include clean, dry, draftless calving quarters, iodine dip of the navel, and colostrum at birth. If you suspect your calf has the disease, call the vet. Because of the severe dehydration, electrolyte solutions may have to be infused to restore the normal fluid balance in the calf. Antibiotics and sulfa drugs may be given. Separate the calf from any others you might have, and clean and disinfect the pen where the cow calved.

Another major calf disease is *common scours*; it is most prevalent during milk feeding, but can occur at any age. The feces are thin and runny, but a normal color. The calf appears droopy and listless, with dull eyes and an elevated temperature and respiration. Scours itself doesn't usually result in death, but it does make the calf more susceptible to infectious diseases (like pneumonia). Often, scours is caused by overfeeding, but poor sanitation, damp quarters, irregular feeding schedules, irregular feeding temperatures, or a rapid change in feeds can also cause the disease. The first thing to do is cut the amount of feed by half, but keep the liquid the same (substitute water for deducted milk) to help prevent dehydration. Reduce milk intake by half for a day, then gradually increase it again. Terramycin can be given to control scours, but if the situation persists, consult your vet.

The third major calf disease and the most common cause of death in calves is *pneumonia*. It frequently occurs three to eight weeks after birth and is usually preceded by some weakening stress on the calf (such as scours or dampness). Symptoms include coughing, raised temperature and respiration, dullness, loss of appetite, general weakened condition, and foul smelling nasal discharge. Although caused by a virus, pneumonia can be the result of cold, damp, poorly ventilated pens, or chills caused by sudden temperature changes. A warm, dry, clean environment for the calf and regular feeding will help prevent the disease. Should pneumonia develop, move the calf to clean, dry quarters. Sulfa drugs and antibiotics may help, but your vet should be consulted.

In general, the occurrence of disease can be minimized by making sure the cow is properly fed before calving to insure the production of a thrifty calf to start with. Colostrum, treated navel, and clean, dry, draftless quarters all have been mentioned. Milk feeding should be kept on schedule, and the amount fed per day should not exceed 10 percent of the calf's weight, with the temperature of the milk and the feeding routine kept constant. Sanitation of calf pens and feeding equipment is extremely important. Sick calves are really pitiful, and losing a calf is terribly sad. Sometimes illnesses can't be avoided, but you should do as much as possible to avoid what diseases and disease conditions you can.

Diseases and Problems

Many of the major problems affecting dairy cattle can be divided into three categories: udder problems (mastitis, udder edema), noninfectious or metabolic diseases (ketosis, milk fever, bloat), and reproductive diseases (brucellosis, leptospirosis, vibriosis). Of course, there are a thousand natural shocks which cattle flesh is heir to—some which do, some which don't fall into these categories—but these are among the more common. Each of these diseases will be covered briefly, more for the sake of making the cow owner aware of danger signals than for treatment. There's a good chance you will never see any of these diseases, and the worst you'll have to deal with will be first aid, but remember that your best friend in the case of a sick animal is the vet. Most large-animal vets are willing to discuss the problem on the phone; they're busy folks, and they don't want to come out to your place unnecessarily any more than you want them to. But although it's always best to get expert advice, it's also a good idea to have some notion of what you're seeing (or what the vet is talking about).

Mastitis is the most common dairy disease in the United States and can affect your herd of one almost as easily as a herd of one hundred. Mastitis, an inflammation of the udder, can be caused by one of at least twenty different bacteria. A cow can have the bacteria in her udder but show no signs of the disease unless there is a lab examination of her milk (nonclinical mastitis). Clinical mastitis is the obvious one: a visible inflammation in the udder or a change in the milk composition. During mastitis, whether clinical or nonclinical, some udder tissue is being destroyed; the extent of the damage depends on the severity of the clinical case.

Why do some cows get mastitis while others don't? It is not really known, but it appears that cows with eroded teat ends, injured teats, or chapped or blemished teats are more susceptible.

Faster milking cows, too, are more prone to the disease, and inheritance may play some part in resistance to the development of clinical mastitis. It has been shown that incidence of infection is highest early in the dry period and early in lactation.

A mild clinical case of mastitis can be detected by the cow owner through routine barn tests: using the strip cup to check for lumpy, ropy, or bloody milk, or using pH cards. The symptoms of severe clinical mastitis are easily observed: pain, redness, swelling, and heat in the infected quarter. Frequently a quarter of the udder will go from nonclinical to mild or severe clinical (a flare-up), and then back to nonclinical after antibiotic treatment. A cow that periodically flares up is said to have chronic mastitis.

Although the organisms that cause mastitis are quite common in the environment and virtually impossible to eliminate from the surroundings, the most usual source of the bacteria is infected cows; the bacteria are transferred from an infected animal to an uninfected one via the milker's hands or the udder washing cloth. The idea in preventing mastitis is to try to eliminate the organisms from the udder. Using individual paper towels for udder washing, dipping teat ends in sanitizing solution after milking, and keeping your own hands clean will help reduce the occurrence of mastitis. Keeping the cows' quarters clean and dry is also of value in controlling the growth and spread of the organism. Certainly don't go milk your friend's cow and then milk your own without washing your hands well first. Treating the cow at the beginning of the dry period also helps control infections; dry cow treatments are available at most farm supply stores. If clinical mastitis occurs during lactation, antibiotics are used for treatment (penicillin or a penicillin–streptomycin combination). Milk from treated quarters cannot be used for thirty-two to ninety-six hours after withdrawal of the antibiotic (it will say exactly how long on the antibiotic bottle, or your vet will tell you), but mastitis milk can be fed to other animals.

There is a school of thought that advocates the use of vinegar in the drinking water to control mastitis. It is said that a tablespoon of vinegar in the water daily will do the trick, and a lot of people say it works. I certainly won't say it doesn't, but I confess to using a teat dip after milking.

Udder edema is a fluid collection causing a swelling of the udder either just before or just at calving, occurring more commonly in first-calf heifers than in olders cows. The accumulation of fluid in the lower udder makes it hard to the touch and uncomfortable for the cow; fluid may also build up just in front of the udder. There is no heat. The tissue fluid accumulates because of an upset in the balance between lymph formation and drainage from the udder, caused by a variety of complex factors (high salt intake, high fluid intake, low tissue pressure—anything that changes the osmotic pressures of the tissue fluids and the blood). Milk produced by cows with udder edema is normal, and in most cases, the edema decreases within two to four days and is gone completely two to three weeks after calving, without treatment. In severe or repeat cases, diuretics may be used; massaging the udder also may be helpful. There is no real way to prevent the occurrence of udder edema.

Ketosis is not usually encountered by the family cow owner because it is primarily a problem with very high-producing cows. The first three weeks after calving is the critical period during which ketosis may develop, although it can occur from ten days to six weeks after calving. The symptoms include going off feed (first grain, then silage, and if severe, hay), a weight loss, a drop in milk production, rumen inactivity and constipation, dullness, listlessness (although sometimes it's the opposite: the cow is nervous, restless, and may kick or butt), and a characteristic acetone odor on the breath. (Sensitive-nosed dairy people can walk into a barn full of cows and pick out the one with ketosis by smell alone.) The symptoms are all associated with the cow going off feed. What causes the cow to go off feed is unknown, but ketosis usually occurs only in high-producing cows on barn feeding; cows on pasture rarely get the disease. A diagnostic test for ketosis involves sampling the urine for the presence of high levels of ketone bodies (the products of rumen action on fatty acids). Many cows recover spontaneously, but treatments include glucose injections, glucocorticoid injections, or propylene glycol or sodium propionate given orally for five to six days (by drench; the cow won't touch the stuff). Again, speak to your vet.

Although ketosis can't be prevented entirely, recommendations include the following: avoid having cows overconditioned at calving, increase concentrates as quickly as possible after calving (but not so quickly as to throw the cow off feed), avoid abrupt changes in the ration, feed good quality hay, provide adequate ventilation in the barn, and make sure the cows get exercise daily.

Bloat is simply severe gas pain. The left flank of the cow protrudes or is extremely swollen because of excessive gas accumulation in the rumen. The animal may stand with her nose thrust forward and her mouth open, trying to get rid of the gas. Death is common if severe bloat is not treated immediately.

Bloat most frequently occurs in dairy cows grazing legume pasture, such as alfalfa and ladino clover; birdsfoot trefoil, however (a legume), will not cause bloat, nor will grass pastures.

Because bloat occurs quickly and must be treated quickly, usually it is a job to do yourself. The first thing to do is to see if you can get the cow to belch; tying a stick in the cow's mouth sometimes works. Try to get the animal to move around. Next, try putting a garden hose (¾-inch diameter) down the throat and into the rumen to release any trapped gas (blow in the tube every now and then to clear it). If nothing happens, pour in 1 to 1½ cups of vegetable oil, or if necessary, cream. If this remedy doesn't work and the animal is in real trouble and lying down, rumenotomy is the absolute last resort: the body and rumen walls are punctured on the left side between the last rib and the hipbone (about 2 to 3 inches down from the hipbone). The best instruments to use are a trocar and cannula, leaving the cannula in place so the gas can escape. If you haven't got these, you'll have to settle for any sharp instrument that will do the job and keep the opening open.

To prevent bloat, use mixed grass–legume pastures. Early, fast-growing plants produce bloat most often, so hold off a bit when the pasture is at this stage. If you're lucky enough to have a lush legume pasture, provide your cow with some dry feed in addition to the grazing to try to cut down on the amount of pasture she'll consume.

Milk fever (parturient paresis) is one disease many family cow owners might see, since its incidence is two to three times higher in Jerseys than in the other breeds. It is more common in high-producing, older cows and tends to repeat (a cow that's had it once has a good chance of having it again). Milk fever occurs at or within two or three days of calving. The symptoms (in order of increasing severity) are loss of appetite, rumen inactivity, dull eyes, cold ears, a stilted gait, and paralysis. A cow's typical milk fever position is lying on her breast or right side with her head tucked back into her flank. Coma and death usually follow unless the animal is treated. If the cow lies in a cramped position for more than an hour or so, she may become unable to get up because of degenerative muscle changes; ruptured ligaments and tendons can result when the cow strains to get up. (The problem is worse in a stanchion or if the cow tries to get up on an unbedded, hard, slippery surface.)

Milk fever is caused by a sudden drop in blood calcium—related to the rush of calcium into the milk at calving—coupled with a malfunction of the body's regulatory mechanisms. The most common treatment is an intravenous injection of calcium gluconate to get calcium back into the blood as quickly as possible; if done quickly and correctly, recovery can be dramatic—a cow that was down can be up and around in an hour. Call your vet.

The major preventative measure is to keep the ratio of calcium to phosphorous in the diet of the dry cow at 2:1—which seems to help keep the regulatory mechanisms regulating. How do you know if it's a 2:1 ratio? You can use the feeding standards to try to figure it out or talk to your vet or county agent about the diet you're feeding.

Brucellosis was on the way out in the Northeast, but is showing up again in dairy herds here and there. The organism that causes brucellosis in cows causes undulant fever in humans and is transmitted through the milk. The major symptom of brucellosis is abortion in mid- or late pregnancy; retained placenta may be a problem and may be followed by temporary or permanent infertility. Annual blood tests can be made (often at the state's expense), and a cow testing positive should be destroyed, as the disease spreads rapidly and there is no known cure.

Vaccination of calves between three and six months old is the usual method of prevention. Some vaccinated calves, however, will show positive tests as adults (even though they're not infected), resulting in some confusion in later diagnostic work. For this reason, some states recommend against vaccination. Your vet can tell you about the status of brucellosis in your area.

Vibriosis is characterized by infertility and abortion. Long, irregular estrous cycles are the most common symptom, along with poor conception rates. If you breed your cow artificially, vibriosis is not of much concern, since it is transmitted by the bull in natural service. (It can be transmitted through the semen in A.I., but A.I. bulls are very carefully tested.) There is no effective treatment for vibriosis, but your vet should be consulted.

Leptospirosis is of concern, even for the one-cow owner, since the parasite causing the disease can be spread not only by cows (through the urine) but also by other ungulants (such as deer) or by rodents. Symptoms include a rise in body temperature and a drop in milk production. The milk may be thick, bloody, or yellowish. The urine may seem reddish. Symptoms may last only about a week, but the cow frequently aborts one to three weeks after the symptoms disappear (usually after the fourth month of pregnancy).

Leptospirosis can be controlled by vaccination, which gives immunity for twelve to eighteen months, and antibiotics may clear up an existing infection. Consult your vet about the need for vaccination in your area, for diagnosis if you suspect the disease, and for treatment.

Milking

After all the talk about feeding, breeding, calving, housing, etc., it seems high time to get down to the nitty gritty of owning a dairy cow: milking.

The easiest way to learn to milk is firsthand from someone who already knows how. If you don't have that opportunity, though, don't be discouraged; you can easily learn on your own cow. If you're lucky you'll be learning on an older animal; she'll be much more tolerant of your fumblings. (I remember when one of our friends was learning to milk on Betsy. Usually we milked her out in about ten minutes, but poor Tom was at it for ages. After an hour we checked back: Tom was almost finished, and Betsy was standing there—asleep! How much more tolerant can you get?)

Cleanliness and efficiency are the overriding factors in all the steps of milking. Although some people pooh pooh the issue of cleanliness, there is no question that clean milk keeps longer, tastes better, and is crucial to successful cheesemaking. A clean and efficient setup will make life easier for all concerned, and thought and preparation in this matter before you get the animal will save a lot of trouble.

A separate space for milking (not the cow's pen) is the first step in the production of clean milk. The floor in the milking area should be such that it can be swept, or even better, washed; a concrete floor answers perfectly but is not absolutely necessary. The more tightly enclosed the space is, the better—tight walls cut down on the circulation of dust and dirt; but adequate ventilation is also important. If possible, the walls should have a washable surface (painted enamel). Having water easily available will help in keeping things clean; you're much more likely to wash up well if you don't have to haul water three miles to do so.

There are probably as many different milking setups and procedures as there are family cow owners, but I can describe ours as one possible way to organize.

Our milking area is a partially enclosed, concrete floored space (with a convenient hole in the floor for a drain), which we have equipped with a table, a drying rack for the milking utensils, a shelf (for things like brushes, filters, medication, etc.), a refrigerator, lights, a place to tie the cow, and a fifty-five-gallon water drum, easily kept filled with cold water. In the morning we carry up a gallon or so of hot water, empty it into a waiting dishpan on the table, and add a dab of sanitizer. The cow's morning grain allowance is dumped into the feed bucket; then we let in the (by now) pawing and snorting beast, who knowing exactly where her every meal comes from, zeros right in on it. A swivel clip is hanging there to hook onto the rope she wears around her neck, just in case she gets any fancy ideas about walking out before we're through. Her flank and udder are brushed, then we wash the udder thoroughly with paper towels dipped in the hot water. The udder is then well dried and so are our hands (milking with wet hands is definitely a no-no —a prime spreader of disease and other problems). The first one or two streams of milk from each teat are milked onto a strip cup (a tin cup with a black, dished lid) for two reasons: (1) the first couple of jets are the most bacteria laden and should not go into the milk pail; (2) it is a simple, quick, daily mastitis check. The strip cup is designed to show up problem milk—lumpy, flaky, stringy, bloody, etc. (The strip cup milk can be given to any waiting dog or cat.) Next, we take our seamless aluminum milk pail (unfortunately not stainless, which is ideal, but prohibitively expensive), which is sitting upside down on the rack, all clean and ready from the last milking, and we milk away. Some detail on just how to "milk away" might be in order here.

First, you have to get close enough and low enough to be comfortable. We use a regular, old-fashioned three-legged milking stool, but previously we used one of those old plastic milk crates you can find around. You pull the seat in really close and clasp the milk pail between your knees. Some people milk with the pail on the floor, but this position can lead to problems with cleanliness and also to a loss of flexibility if you have to make a quick move. Also, with the bucket on the floor, it's a lot easier for the cow to step in it or kick it. (I'll bet cows prefer the bucket-on-the-floor technique.) Next—getting the milk out of the udder. Now, right now while you're reading, open your hand and then slowly close it, observing your fingers. You'll notice that you close your little finger first, then the ring finger, etc.; this is the natural and automatic way to close your hand. Now try closing it in reverse—index finger first, then middle, etc. Unless you've been milking lately, you'll probably find this way somewhat harder. This reverse closing is the motion of milking.

Okay, you've got the action. Now, the idea is that the storage tank in the teat is full of milk and more will flow in as soon as you empty out what's there. With your thumb and index finger, squeeze the top of the teat to close it off from the rest of the udder (to prevent squeezing milk back up into the udder, a self-defeating system if ever there was one). After the teat is closed off, use the downward squeeze motion (you don't have to, and shouldn't, pull), and you should get—nothing. Don't be disappointed, most people get nothing the first time they try. Just open your hand completely, and try again; be sure to open up the top of the teat to allow the milk to flow into it. Getting the motion rhythmic and with both hands takes some practice and patience. So does aim; expect some damp shoes, damp pants, a lot of laughs, and not much milk those first few tries. Remember, though: let the milk into the teat, close if off, squeeze it out, let the milk in again, close it off, etc. Milking is a wonderfully soothing and relaxing process, and after a few weeks, you'll be amazed that you ever had to think about the mechanics of it. (Instead of milking and thinking "close off, squeeze out," you can think of much more exalted things, like "the toilet paper in the bathroom needs replacing" and so forth.)

But, as milking is soothing for you, so should it be stressless and soothing for the cow. There is, actually, a physiological reason: the famous milk let-down, without which you cannot milk. When the cow receives the milking stimulus—a calf bumping the udder, the warm udder wash, or just the premilking sights and sounds—a hormone, oxytocin, is released. Oxytocin is the milk-releasing hormone which acts on the muscles surrounding the milk-producing cells, squeezing the milk into the storage compartments of the udder: the milk let-down. This let-down enables you or the nursing animal to extract the milk. If, however, the animal gets frightened or upset by yelling, hitting, etc., another hormone, adrenalin, is released, which blocks the action of the oxytocin, and there is no milk let-down. Actually, milk let-down is quite easy to see. The cow comes into the milk area, you wash her quickly, and the teats just don't seem to have any milk in them. Well, maybe the stimulus wasn't enough. Try washing again, massaging the udder as you do so. With the proper stimulus, the milk will flow into the teats, and they will visibly swell. Immediately after a good let-down, milking is sort of like squeezing water balloons.

After the cow is milked out completely (which she should be at each milking, or you'll encourage mastitis and the cow will cut back on her production), we dip her teats in a solution mixed from the sanitizer. This is a simple step to prevent mastitis and should be done after each milking. Now the cow's job is done, and she can go back to her main business in life: eating hay or grazing.

The milk should now be strained through a filter into a container and cooled rapidly. For straining the milk, some people use a flannel cloth over a glass jar—a very poor practice in terms of cleanliness and labor efficiency. We use a seamless aluminum strainer with disposable milk filters. You can rig up a strainer of your own design, but the material used should be seamless (bugs love to gather in hard-to-clean seams) and should not be plastic. After the milk has been filtered, we pop it into the freezer or refrigerator, or you can immerse the container in cold water; the idea is to cool it rapidly to slow bacterial growth. The milk comes out at over 100 degrees, and commercial dairies aim for 40 degrees within two hours. The milking utensils and strainer are then soaked for three to four minutes in the hot, sanitized water still patiently waiting in the dishpan and then scrubbed with a brush. They're put aside down on the rack to allow them to air dry; do not wipe milking utensils dry. The table is wiped off, the water spilled out, and the milking area swept or rinsed down (depending on how ornery the beast was). Everything is then ready and clean for the next milking. As a final step, we note down how much milk we got at each milking, which helps in determining how much to feed, in heat detection, and in economic calculations.

A couple of problems you might run into are yield fluctuations or difficult milking and off-flavors in milk. The section on lactation discusses yield fluctuations, but if the cow seems difficult to handle and erratic in her yields, be sure that you're milking at the same time every day and keeping the same routines. Maintaining a set schedule is important; don't forget to milk one evening or milk at 7:00 one morning and 10:00 the next. Off-flavors can be caused by a variety of factors but are most commonly caused by unsanitary conditions. If the milk tastes "barny" or "cowy," check your ventilation. Avoid having someone sweep or feed hay while you're milking. If milk sours quickly, improve your milk handling procedures with regard to sanitation. Be sure to brush the cow's udder and flank before milking to lessen "fallout" into the milk pail. Weird feeds can also cause off-flavors, the most often cited being wild onion causing onion milk. If your cow is grazing weedy pasture and possibly eating strange-flavored things, take her off pasture four hours before milking. This same effect can be the result of some hays ingested within four hours of milking; alfalfa, particularly, can sometimes cause off-flavors.

When milking a cow, there is, of course, the question of pasteurization. The necessity for pasteurization of your own animal's milk is debatable. The major problems are those organisms that cause brucellosis and tuberculosis in the cow or undulant fever and tuberculosis in humans, although there are other organisms milk can carry that can cause more minor problems. Many states will pay for your cow to be tested for brucellosis and tuberculosis every three years, but raw-milk drinkers might choose to have their cows tested every year. Because pasteurization kills many of the bacteria in milk, it also improves its keeping qualities. However, on the other side of the coin, one could say that keeping quality is not particularly important to the family cow owner; unlike supermarkets where milk might be two weeks removed from the cow, most people use their milk within two or three days. If the animal is clean, the utensils sanitized, the milking area clean and well ventilated, the milk rapidly cooled and used, and the cow tested, I don't think pasteurization is necessary. However, this is a personal decision milk drinkers must make for themselves.

Dairy Products

No self-respecting dairy cow owner goes for long without making butter. Maintaining a high quality, distinctive tasting butter is an art but don't let this scare you: making perfectly acceptable (even good) butter is pretty easy. Skim the cream from the milk after allowing the milk to sit undisturbed for eight to twelve hours. If you're dealing with a small quantity of cream at any one time (like a quart), the simplest way to make butter is to put the cream in a jar and shake it. With any butter-making technique, it's best to let the cream come to about sixty degrees before churning. The jar should not be more than half full, so either split the cream between two jars or use a big jar. Then shake. And shake. And shake some more. Depending on the heaviness of the cream and its temperature, in five to ten minutes it will get very thick and possibly very hard to shake. Hang in there. In just a few more minutes, clumps will form and start to pull together, and then bingo—the butter will separate from the buttermilk. You can't miss it—you never have to wonder, "Is it butter yet?" Yellow lumps about the size of wheat grains will be floating in bluish milk. A few more shakes will glue all the grains together, and you'll have one big yellow lump. That's it.

Pour off the buttermilk. You can save it and drink it, cook with it, or feed it to pigs, dogs, calves, etc. Beware, though—this is not the same buttermilk you buy in the supermarket. Commercial buttermilk is cultured, making it thick and yogurty. This buttermilk tastes much more like plain milk—just super skimmed.

The lump of butter you're left with still has a lot of buttermilk in it which must be removed. This is one area where art comes into the matter. The first step is rinsing. Put the butter into a bowl, pour very cold (preferably ice) water over it, and sort of slice it over and over with the edge of a paddle, essentially exposing as much surface area to the water as possible. Pour the water off; then repeat the process until the water is clear (the buttermilk clouds it). Butter is not supposed to be rinsed more than three times; overrinsing is said to give it a flat taste. If you want salted butter, now is the time to work in the salt. After the butter has been rinsed, sprinkle on about a teaspoon of salt per pound of butter (or as much as personal taste dictates), and work it in gradually, using the slicing action. Adding salt will help extract more buttermilk.

The next step is to squeeze out the remaining water—again, a developed technique. Basically, you take a wooden paddle and press the butter into shape, at the same time pressing out excess water. Overworking the butter, however, or smearing it along the side of the bowl will cause a greasy texture (but one you can live with). Once the water is worked out, wrap the butter in waxed paper and refrigerate or freeze it.

For larger quantities of cream, the technique is the same once you have the butter separated from the buttermilk; it's getting to that point that is slightly different. Butter churns come in a variety of types and sizes. There's the big old-fashioned kind of tub with the stick and paddle that gets moved up and down; that churn is for a quantity of cream I've never had to deal with. For the person churning a couple of quarts to a gallon or so, there are medium-sized hand-operated or electric churns. The hand-operated ones are glass jars with a screw cap attached to a paddle on one end and a geared crank on top. The electric one is pretty much the same; instead of a hand crank, though, there's a small motor. With either type, the jars should be extremely clean (I rinse it with scalding water before each use) and filled not more than two-thirds full. Yield depends on the heaviness of the cream (and how sloppy you were skimming it), but as a rough guide, a quart of cream should yield half to three-quarters of a pound of butter. It should be mentioned, by the way, that some blenders will churn butter.

Where's more room for developing the art? Well, for one thing, in judging the age and ripeness of the cream. Fresh cream should not be churned; it should be at least a day old. There are those who claim that the best butter is made from cream that is actually sour. I tried it once; it was good, but I don't know if my palate is really refined enough to discern subtle differences in butter. In any event, I did discover that if cream went sour, it could still be made into perfectly good butter. It is also said that cream from different days should not be mixed together into one batch. However, I always mix cream from different days, and it never caused any problems. The temperature of the cream and of the room also will affect the quality of the butter. Churning time is another factor: overchurned butter is hard; underchurned is soft, gooey, and difficult to work. As already mentioned, the amount of rinsing and amount and style of working will influence the taste and texture of the butter. Obviously, the degree of salting will affect the taste and will also affect the keeping quality.

Butter color is affected by a few things. Much commercial butter is artificially colored, and you can color your homemade butter, too. But when you have a dairy cow, it's kind of fun to watch the color of the butter change with the seasons. Early spring and summer butter is the richest, most gorgeous yellow you've ever seen—it looks fake. Winter butter can be almost white. The change is, of course, a result of the cow's diet. The lush spring grass is full of carotene (Vitamin A), which gives the yellowness. Winter roughages are low in carotene—hence the paling. (You want to see yellow butter; try making butter right after the cow stops producing colostrum—it's amazing.) In addition to diet, the breed of the animal will also affect color; Jerseys and Guernseys in particular are noted for their yellow fat.

Diet of the animal can also play a role in the quality of the fat; timothy hay, for example, could cause a cream that is difficult to churn. Strange weeds or plants could cause off-flavors. Other factors affecting the flavor of the butter are cleanliness (of hands, utensils, churn) and, oddly enough, the bowl you use. Most people use a wooden bowl (I can't imagine anything else), which will sometimes absorb flavors of other things that may have been put in it; it's a good idea to keep one bowl specifically for butter. Apparently, too, some woods impart their own woody taste to the butter.

However, in spite of all the areas for improvement in making butter, there really aren't that many areas in which to fail. In the several years that I've been making butter, I never had any that was unusable. Some butters weren't as good as others, for sure, but they were all edible. So don't be intimidated by "this affects that" and long directions; it's easy, satisfying, and economical.

Cheesemaking, now, is a whole other ballgame since most cheese recipes are at least time consuming and at most extremely complicated. It is, however, a field in which one can work and improve for years, producing truly superior and interesting products. It takes time, devotion, a lot of experimentation, and unlike buttermaking, with cheesemaking you have to expect a lot of failures. (A good time to start learning is when you have pigs handy. They never consider *anything* a failure, and at least you can console yourself that your effort wasn't a total waste.)

The books listed in the References present a good starting place; the cream cheese in the Minnesota Extension Bulletin and Hobson's mozzarella are especially easy for beginners.

But definitely give cheese and buttermaking a whirl. After all, that's why your dairy queen is around!

References

Dairy Cattle

BECKER, R. B. *Dairy Cattle Breeds*. Gainesville: University of Florida Press, 1973.

NORDBY, J. E., and H. E. LATTIG. *Selecting, Fitting, and Showing Dairy Cattle*. Danville, Ill.: The Interestate Printers and Publishers, 1961.

REAVES, P. M., and H. O. HENDERSON. *Dairy Cattle Feeding and Management*. New York: John Wiley and Sons, Inc., 1963.

SCHMIDT, G. H., and L. D. VAN VLECK. *Principles of Dairy Science*. San Francisco: W. H. Freeman and Co., 1974.

TRIMBERGER, G. H. *Dairy Cattle Judging Techniques*. Englewood Cliffs, N.J.: Prentice-Hall, Inc.

VAN LOON, D. *The Family Cow*. Charlotte, Vt.: Garden Way Publishing, 1976.

Dairy Products

HOBSON, P. *Making Homemade Cheeses and Butter*. Charlotte, Vt.: Garden Way Publishing, 1973.

ZOTTOLA, E. A., and H. A. MORRIS. *Making Cheese at Home*. Agricultural Extension Service Bulletin 395. Minneapolis: University of Minnesota, 1975.

4
DAIRY GOATS

Without doubt, first prize for dairy world underdog goes to the goat; among mammals in general, the goat is probably the only tough competition for the wolf for most maligned. Reduced to a common scornful epithet ("you old goat"), even in a comic strip (Charlie Brown's "I could be the hero . . . or the goat"), the goat has a lot of bad public relations to combat. As is often the case, this image is not justified; on the contrary, the goat, although assuming a minor role in U.S. agriculture, is a major component of the dairy industry in many parts of the world. Numerous virtues of the dairy goat make her valuable in third world and Mediterranean countries, and many of these same virtues can be capitalized on by the small stockholder here in the United States.

Of prime consideration is the grazing ability of the goat. Although a ruminant like the cow and sheep, the goat stands out in the quantity of feed it can consume for its size: 6.5 to 11 percent of its body weight in dry matter, as compared to 2.5 to 3 percent for sheep and cattle. In spite of the fact that each 100 pounds of goat requires 1.5 times as much feed for maintenance as does each 100 pounds of cow, because the goat can consume two to three times as much feed as the cow, more feed nutrients remain available for milk production. What this boils down to is that a goat can survive, and produce milk, on grazing that couldn't support a cow, largely because of the goat's ability to eat more. Additionally, goats will consume with delight roughages a cow wouldn't touch (and possibly couldn't touch; goats are much more adaptable to rough terrain): bushes, brambles, weeds, thistle, twigs, etc., collectively known as browse. When dealing with improved pastures, this advantage is greatly diminished, but it can easily be seen that in many parts of the world (and in many backyards), improved pastures are impossible—hence, the supremacy of the goat.

Special qualities attributed to goat's milk, which differs in composition from cow's milk, are important. First is its comparative ease of digestibility, largely because of the smaller size of the fat globules, which are 2 microns compared to 2.5 to 3.5 microns in cow's milk. For this reason, goat's milk is often prescribed for people with ulcers or other digestive disorders. Second, many children as well as some adults are allergic to cow's milk but not to goat's milk. Also, the latter is somewhat lower in lactose, so in areas where people are lactose-intolerant (which is upwards of 50 percent of the population in some African and Asian communities), goat's milk poses less of a problem. A few other points in its favor are a mild laxative effect, a higher buffering quality, a higher phosphate content, and a higher Vitamin B content.

In any event, be aware that goats are not the smelly, tin-can eating scavengers they're made out to be; they are practical little dairy animals deserving recognition of the honorable place they hold in the dairy world. Far from being a poor man's cow, the goat

Fig. 4-1. Parts of a Dairy Goat

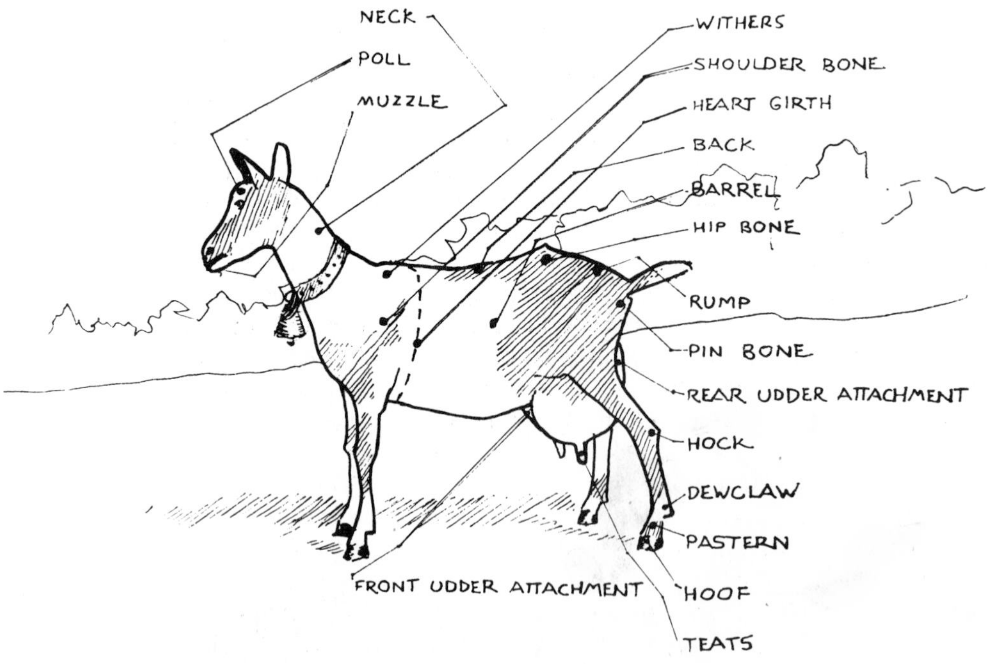

fills a gap in the needs of many that a cow could never fill; wherever land is poor, water and grazing limited, or small quantities of milk desired, the dairy goat fits right in. Neither, however, are goats miniature cows, so attention must be paid to their special requirements.

Breeds

There are five major breeds of dairy goat in the United States.

Alpines are usually of the French Alpine variety, although there also exist (less commonly) Rock Alpines and Swiss Alpines. The French Alpine varies in color from white, gray, brown, and black, to red and combinations of these colors. It's a large, rugged animal, hardy and with few kidding problems. The average doe weighs around 125 pounds and is known as a good, steady, high producer (recent breed record: 4,826 pounds in a 305-day lactation).

Nubians are known as the Jerseys of the goat world because although they produce less milk than the other breeds, it is high in butterfat (breed record: 4,392 pounds). Originating in Africa, the Nubian has a short, sleek coat, usually black and tan, a Roman nose, and long, drooping ears. A very gentle breed, they thrive on companionship and can be remarkably persistent in their demands

Toggenburg.
Courtesy of M. Wood, Cornell University Extension Service, N.Y.

for attention. The average mature doe weighs around 130 pounds.

The *Toggenburg,* a Swiss breed, is a somewhat smaller goat, with the average doe weighing about 100 pounds. The color characteristic of the Togg is the light stripe down each side of the face, white lower legs, and a body color some shade of brown. They may be long- or short-haired; for dairy purposes, the short-haired are more desirable. Toggs are good milkers, with the breed record of 4,750 pounds.

If Nubians are the Jerseys of the goat world, *Saanens* are the Holsteins. A white, good-sized goat (does at 135 pounds), originating in Switzerland, the Saanen gives large quantities of less rich milk (breed record: 4,905 pounds). They are placid animals and seem to be better suited to a more confined situation than the other breeds.

LaManchas are a newly developed breed, with Spanish ancestry. They are considered extremely hardy and adaptable and produce milk that is fairly high in butterfat. The most noticeable feature of the LaMancha is its lack of ears (which may sound dreadful but is really amazingly cute).

In addition to these purebred goat classifications, there also are recorded grades, grades, and scrubs. A recorded grade has one registered purebred parent and the other of unknown origin. If both sire and dam are registered purebreds, but of different breeds, the kid is a recorded grade and is classified as experimental. Offspring of recorded does can be upgraded by continued mating to a purebred sire of one breed. By the third generation, the kids are seven-eighths purebred and can be registered as purebreds. A plain grade goat is one without registered parents (known) but showing some definite breed characteristics. The grade goat is still several steps above the

Alpine.
Courtesy of M. Wood, Cornell University Extension Service, N.Y.

Nubian.
Courtesy of M. Wood, Cornell University Extension Service, N.Y.

scrub, a goat with few clues to its parentage, your basic 57 varieties mutt.

Selecting a Dairy Goat

For the beginner, selecting a goat can be an awesome task. First, consider whether you want a purebred; unless you are planning to show or to go into breeding, however, for backyard milking a grade doe would probably be best. Local 4-H Clubs or extension offices can usually put you in touch with goat owners in your area. From that point on, selecting a dairy goat is much like selecting a cow as far as the dairy characteristics are concerned.

As with a dairy cow, in addition to overall health, appearance, and dairy character, the udder should be closely examined. Teats (only two) should be of good size, uniform, free from injury, squarely placed, and distinct from the rest of the udder. The udder should be capacious, well attached, and evenly balanced. Again, an inexperienced milker should try to acquire an experienced goat who has passed close scrutiny at milking time.

Milk production is of prime importance for the buyer wanting one or two goats to meet a family's needs. A good goat produces about three quarts per day (a very good one will give a gallon), and lactation length should not have been less than seven months. Grade goats meeting these production criteria should not be difficult to come by; unfortunately, poorly producing, badly kept scrub goats are easier to come by. But remain firm, and don't settle for a less-than-good goat, be it grade, scrub, or purebred.

Saanen.
Courtesy of M. Wood, Cornell University Extension Service, N.Y.

LaMancha.
Courtesy of M. Wood, Cornell University Extension Service, N.Y.

Management

Housing and Equipment

Housing for goats need not be elaborate, but it must offer protection, be dry and draft-free, and have good light and ventilation. Anything from a pen in the corner of an existing barn or garage to a shed constructed specifically for your goats can serve this purpose. Even more than other livestock, though, goats are particularly sensitive to drafts and dampness, so pay careful attention to these factors when selecting your pen site. Pen space per goat should be a *minimum* of twenty-five square feet (five feet square or four by six feet, etc.), but you mustn't overlook other needs, such as kidding space, feed storage, and milking area. (A kidding pen can double for a hay storage area when not needed for kidding. Why have thirty square feet sitting empty waiting to be used only once a year?) Four-foot high walls should keep the doe in. If you have more than one doe, they can be housed together in a loose housing arrangement, sharing a communal hayrack and waterer. They should not, however, be forced to share grain; they won't do it, and the head goat will get it all. Feeding grain to the does while milking avoids this problem.

Contrary to the popular image, goats are extremely fussy about what they eat and drink and will not touch anything even slightly contaminated. A goat will pull all the hay out of the rack onto the floor, looking for just the right bit, and then refuse all the rest because it's been stepped on. In the process a few hay stems might have fallen into the water bucket—now the goat won't drink the water. And so, the challenge is born: how to design a hayrack and a waterer that will eliminate the problem of wasted hay and ever-changing water buckets. One solution to the water bucket problem is to cut a goat-head-sized hole in the front wall of the pen and hang the water bucket on the *outside* of the pen; the goat sticks her head through the hole to drink, eliminating nanny berries and other unsavories from the water.

A hayrack is more tricky and requires creativity. The rack must keep the hay off the floor and must not be climbable, in or on. A wooden rack, V-shaped in cross section, with slats 1½ inches apart, can be set with the lower part of the rack at the goat's normal eating level and the upper section above it. This design keeps at least some of the hay out of the goat's reach; the top hay will slide down as the bottom hay is (we hope) consumed. Remember that whatever the design, the rack will also serve as leaning post and back scratcher, so a flimsy structure simply will not survive. Hay nets, by the way, are not acceptable, as goats tend to get hung up in them.

The best flooring to use for goats' housing is as debatable as

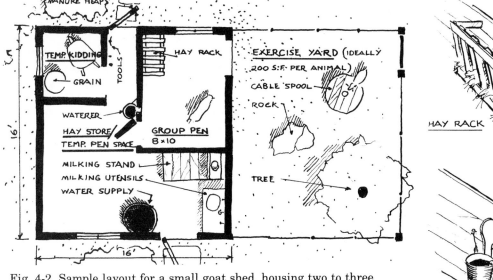

Fig. 4-2. Sample layout for a small goat shed, housing two to three does. A keyhole waterer, with the bucket hung outside the pen, prevents contamination of the water.

that for any livestock. Many people advocate a dirt floor with bedding; moisture (urine) goes through the bedding and drains in the earth, leaving fairly dry bedding on top. The wettest portions of the bedding can be removed and dry added on top, thereby building up a dry, warm, comfortable pack for the goats. Some people are horrified at something as primitive as dirt, however, and choose the more cleanable concrete or wood. As mentioned in Chapter 3, concrete tends to be hard on feet and legs and to be quite cold. It also, obviously, is not porous, therefore requiring large amounts of bedding. Wood is somewhat better, being softer and warmer, but it does absorb odors easily and needs frequent cleaning. This issue is simpler for goats than for cows, however, because goat manure is so much easier to deal with. There's a lot less of it, it's drier, it's lighter, and it can go directly into your garden without needing to be rotted first.

An extra hint for goat comfort in a cold or poorly bedded pen: a sleeping box. A large box, raised six to twelve inches off the floor, will provide a nice, cozy sleeping area for your goat. Just be sure to locate the box where the goat cannot jump to freedom from its top.

A final note on the pen: goats are intelligent and agile and will escape if there is a way. Place gate hooks or latches on the outside of the door low enough so the goat cannot possibly reach them; otherwise she'll figure out in no time that all you have to do is lift this and push that and . . . freedom.

If your goats are not on pasture, they must have an exercise yard. Two hundred square feet, sturdily fenced and complete with toys, should keep a goat happy. Remember that goats are climbers; provide boxes or sturdy crates for the purpose (but not too near the fence). A cable spool is great; it provides a sunny sleeping spot on top, a shady one below, and a place to play king of the mountain. A very large rock or two will provide more climbing area and will help keep hooves trimmed. Seesaws are appreciated and easily made. An empty barrel or nail keg makes a good butting toy. If possible, arrange things so the goat can have free access to the yard from her pen. If this is not possible, be sure that the yard is supplied with water, shade, and a windbreak. If the goat is on pasture, be sure that there, too, water, shade, and windbreak are available.

Fencing

Fencing is the eternal problem of the goat owner. Not only are goats excellent jumpers, but also they are prodigious scratchers and leaners and will quickly reduce a loosely stretched wire fence to a useless, sagging heap. Woven wire, four feet high, tightly stretched and well braced, will usually do the job, but it is awfully expensive to fence in a whole field this way. Barbed wire is a poor choice; it simply presents too many hazards.

Electric fence is by far the cheapest, and it does seem to work most of the time. Some goats can be kept in by a single strand, whereas others require three, with the top strand at 4½ to 5 feet. If you teach your goats about the fence in a small area first—and it's a good, strong lesson—you should be able to get by with two strands. Electric fence must be properly maintained, however, to keep it from shorting out; this takes a certain amount of diligence. Also, don't turn off the current and then touch the fence where the goats can see you; they'll learn that the fence is not always on, and it will be the beginning of the end.

Tethering a goat is a solution only for short, supervised spells. Never leave a goat unattended on a tether; she can tangle herself badly and is a sitting duck ("a tethered goat" is just as good an expression) for dogs. The most gentle of neighborhood dogs can be a serious danger to a goat. Dogs and goats just don't seem to mix, so don't stray out of earshot if your goat is tethered.

Odd as it may look, goats can be taught to walk on a leash like a big dog. In suburban areas or where the goat is confined all day, browsing walks can be a great treat.

Feeding Does

Roughages As always, when feeding a ruminant the first consideration is for the roughages. Of course, better grade hays are the most desirable (leafy green legumes). Poorer hays (grass hays, coarse stemmed, weedy, bleached) can be nutritionally balanced by the concentrate mix, but although the goat will eat the stuff, much of it will be rejected. Green or succulent feeds (silage or root crops) can be fed as part of the roughage ration.

Generally hay is fed free choice, and consumption will amount to two to three pounds per day. Remember the large capacity of the goat: five to seven pounds of dry matter per day, at least half of which must come from roughages.

Pasture is an excellent roughage for the doe. For a milking doe, good pasture can replace one-half the concentrate mix. Dry does can be maintained on pasture and browse alone if they are plentiful and of good quality. With poorer or limited pastures, feed dry does one pound of concentrate per day.

Concentrates For most goat owners, commercial dairy feeds will conveniently and economically supply the concentrate portion of the ration. High-producing does fed poor grass or mixed grass-legume hay need a 16 percent protein ration. Lower-producing does or those with access to good legume hay can be fed a 14 percent mix. Concentrates must be fed to the does individually, with the quantity based on the level of production. For a goat giving up to two quarts of milk per day, one pound of grain per day is sufficient. With production over two quarts, increase the grain allowance by one-half pound for every additional quart produced.

Remember that these rules are not rigid. Be aware of the condition of your animals and the quality and quantity of roughage consumed, and make adjustments in the ration accordingly. If a good milker is getting only fair hay, the concentrate quantities may be increased up to 25 percent. If a doe is on excellent pasture, on the other hand, concentrates may be reduced by half.

Dry pregnant does should have free access to hay, plus one pound of grain per day. Excellent pasture can substitute for the grain.

Water must be provided at all times and must be kept clean. Mineralized salt blocks should also be accessible at all times.

Reproduction

Breeding Breeding a doe is a little more complicated than breeding a cow; A.I. is not yet readily available for goats, and goats are seasonal breeders. Therefore, you must find a buck and get your doe to him at the right time.

August through January is the usual breeding season, and September, October, and November the prime times. During the season, does come into heat approximately every twenty-one days, with the heat period lasting two to three days. If you know your doe fairly well, estrus detection should not be difficult; she'll show a slightly swollen vulva, possibly with a mucous discharge, be restless and uneasy, wag her tail a lot, urinate frequently, and bleat till you go nuts. On the second day of the heat period, the doe is bred, preferably twice. (Once you bring your doe to the buck, it'll be a cinch to see whether you guessed the timing right.) If the doe does not return to heat, it can be assumed that she has been successfully bred.

Although A.I. is not widely practiced for goats, some work is being done along those lines. Information may be available through your local extension office, or you could write for information to Dr. Harry Heiman, Executive Secretary, National Association of Animal Breeders, Columbia, Missouri.

Lactation The lactation period of the doe is normally seven to ten months, allowing for a two-month dry period before kidding. Drying off a goat is much the same as drying off a cow: restrict access to water for a day, reduce grain intake, and stop milking. One milk-out may be needed after several days, but that one milking should suffice (see Chapter 3).

Some extremely milky does will run through a breeding season; that is, instead of kidding once a year, they kid only once every other year, running their lactation eighteen to twenty-two months between kiddings. If your goat is one who maintains a strong lactation and can run through, and you're not particularly interested in the kids, by all means take advantage of the goat's ability.

Kidding Gestation in goats is 151 ±3 days; add 5 months to the breeding date, with a little leeway on each side, and you've got the due date.

The kidding pen can be any clean, draft-free, dry space approximately five or six feet square (thirty square feet). Walls should be solid rather than slatted; as the doe strains during delivery, she often braces her head or legs, so there should be nothing for her to get caught on. No water bucket is necessary, and in fact, one should *not* be placed in the pen; it is not unheard of for newborn kids to drown in a water bucket. Be sure the bedding is clean and dry and that the doe is used to the space (put her in the pen several hours before the event). Signs of approaching kidding are restlessness, bleating, and seeking out a quiet spot. A filling udder (bagging up) is definitely a sign, but it can happen as much as a week before kidding; the doe may need a prepartum milking if this condition is severe. The flanks show a pronounced hollow, and the tail seems high as the ligaments on either side of it loosen and relax. A thick, opaque discharge may be present.

Goats usually have easy kiddings: the doe strains, the water bag appears and breaks, within fifteen minutes the front feet appear, then the nose, and within a half hour the kid is born. More often than not, does have twins or triplets, with rest periods between each appearance. Problems in kidding are like those in calving (see the discussion of abnormal presentations in Chapter 3) except that if you do have to go into a goat, you will have to carefully sort out which head and legs go with whose body before making any decisions or rearrangements. Suspect problems if there is excessive delay (over half an hour) after the water bag breaks.

As soon as each kid is born, its nose should be wiped clear of mucous and its navel dipped in iodine. The afterbirth should then be expelled within four hours; if it is not, call the vet. There may be a watery, bloody discharge for a few days after kidding, which is normal, but if the discharge becomes thick or bright red, consult your vet.

If you want to be extra nice to mama after her labors, warm water with a cup or two of molasses will be welcome, as will a bran mash. Supply the doe with all the hay she'd like and water, but wait to start her grain feeding until the following day; then increase her feed gradually to reach her full ration after several days.

Raising Kids

Raising kids is one of those things for which you develop a feel and style all your own. Read six books on the subject and you'll get six variations on how to be a foster mother.

The first step is to ask yourself, "Do I want the milk from the goat right away?" If the answer is yes, then the solution of leaving the kids with the doe and letting nature take its course is clearly not practical. So when do you separate the kiddies from their mama? Some people advocate immediate separation; don't let the kids nurse at all, and remove them from the doe completely (out of sight and hearing). Then milk the doe out and feed the kids from bottles with lamb nipples. After the colostrum clears up (three to five days), the milk is suitable for human consumption, so you switch the kids to milk replacer, or if the doe is producing enough, continue them on goat's milk. It is critical, however, that the kids get the colostrum within a few hours of birth. A variation on this theme is to separate the kids and feed them from bottles, but then let them return to their mother after a few feedings. The theory here is that the kids will look to *you* and the bottle as their food source, not the udder, so they can have mama's care and company but not her milk.

Yet another system is to leave the kids to nurse the doe until the colostrum has cleared up, then to separate the kids and doe completely, feeding the kids from bottles. Or after a few days, switch the kids to bottles, but separate doe and kids only by a slatted wall so they can still hear, see, sniff, and lick each other. As you can see, there are endless possibilities and variations, and so far, no one has shown that any one method is superior to any other. You'll just have to evolve a system that works well with your doe, your needs, and your particular setup.

When bottle-feeding kids, use rigid bottles and lamb nipples rather than plastic bottles and baby nipples; goats' mouths are more like lambs' mouths than like people's mouths, and the hard-sucking little goat won't collapse a hard bottle the way it would a plastic one. While feeding the kid, you might want to hold the bottle in such a way that a thumb is kept on the nipple so that if the nipple is sucked off (which is not uncommon) it can be retrieved before disappearing down the kid's throat. Holding bottles for more than two kids can be a real treat, by the way. You've got to feed them all at the same time, so how do you do it? You'll find that your knees and feet (in addition to your hands) can hold bottles in ways never before imagined, or you could take the coward's way (and the sensible way) out and buy a gadget called a lamb bar, which is a pail with several nipples coming out of it. Individual nipple pails can also be used.

Milk should be warmed to body temperature and fed three times a day for the first two or three weeks. Some goat raisers will feed four or five times a day for the first several days. Certainly this is more imitative of the natural situation—several small feedings rather than a few big ones—and it is better, but it is also a lot more work. Kids apparently do adequately on three feedings but will be less prone to scour with more; again, it's entirely up to you.

The kid should have sixteen to thirty-two ounces of milk per day, spread over however many feedings you're doing, for the first few weeks, starting initially with somewhat less (approximately sixteen ounces) and only gradually building up to one quart. It is always better to slightly underfeed those first few days than to overfeed, so take it slow and see how the kids handle the quantity of milk given before offering more. After three weeks, gradually increase the daily milk allowance to 1 to 1½ quarts as the kids can handle it. The number of feedings can also be reduced to two. It cannot be emphasized enough how much milk feeding at this early stage is a matter of judgment and sense. Small kids will naturally drink less, and should get less, than big, vigorous kids, and adjustments will have to be made all the time, based on how the kids do. Look out for scouring, watch the kids' behavior, and be sensitive to their weight (which should be gaining, of course); use your eyes and your head rather than a rule book, and things should go well.

By the time the kids are three weeks old, they should have access to all the creep feed they'll eat; a calf starter mix is fine for kids. Good hay, as well as water and salt, should also be available. Depending on how well they're eating grain, the milk feedings can be cut back and eventually eliminated by six weeks to three months. Ideally, the aim is to get the kids on grain and off milk as soon as possible. Weaning can be accomplished by gradually reducing the quantity of milk, cutting out one feeding, or watering the milk more and more each day. Most goat raisers feel that cutting to one milk feeding from two, then no milk feedings at all, is the best way.

Needless to say (but I'll say it anyway), housing for the kids must be kept clean and dry. Exercise space is essential, not only for the sake of the kids but for your own sake as well; there's nothing quite like watching a yardful of playful goat kids.

Once the kids are weaned, you are faced with growing does; feed them one pound of high protein feed per day, plus good hay ad lib (and water and salt, of course). The age at which you should eventually breed your doe depends on her growth. A well-grown doe can be bred during her first fall, if and only if an excellent plane of nutrition is maintained during her pregnancy and lactation since she will still be growing herself as well as producing a kid and milk. More commonly, does are bred at approximately eighteen months. (Goats reach their prime at four to six years but can milk well up to twelve.)

Dehorning of kids is usual and should be done quite early, when the kids are two to five days old, if possible. You can do the job yourself or have a vet take care of it for you. The methods are the same as for disbudding calves: caustic paste or a disbudding

Fig. 4-3. Placing the kid in a dehorning box makes the dehorning job simpler.

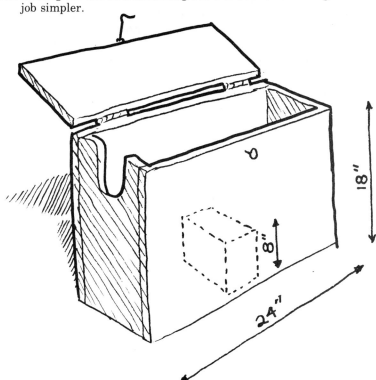

Fig. 4-4. Pattern of hair growth indicating polled (above) or horned (below) condition.

iron. You may be lucky, however, and have hornless kids; the growth pattern of hair on the top of the head can indicate the presence or absence of horns.

So much for doe kids; now what about the other half of the goat population? If you have a buck kid, there are several paths you might follow. Unless the goat has a terrific background, it doesn't make a lot of sense to raise him as an intact male. Some people allow the buck kid the colostrum and then slaughter him for meat. Others get rid of buck kids at birth, and still others raise bucks to keep or sell as pets. More often, bucks are slaughtered for meat once they've reached weaning age. If the buck is going to be around for more than a few days, however, he must be castrated, and the earlier the better; a buck kid is capable of breeding does by the time he is three months old. Again, this procedure can be done by the vet or you can do it yourself. The methods are the same as for calves.

Diseases and Problems

As with any class of livestock (or humans, for that matter), prevention of disease is the method of choice. Good nutrition, clean, well-maintained living quarters, and regular care should greatly reduce the incidence of disease among your goats. Among routine preventative measures are hoof trimming, regular worming, dusting for external parasites, yearly tetanus shots, and enterotoxemia shots before kidding. Consult your vet for any other preventative vaccinations that may be recommended in your area. Daily observation of the animal's behavior and routine checking of the milk will give indications of problems. Any goat acting sick or strange should be isolated, kept warm and dry, and observed closely. A first-aid kit should always be kept handy.

The following are some of the problems you might (but probably won't) run across.

Internal Parasites

This is the most common goat problem, one you *will* run across, and it should be handled routinely once a year. Signs of worm infestation are pale membranes around the mouth and eyes, a rough coat, soft droppings, and thinness. Severe infestations can lead to anemia and damage to the stomach and intestinal walls, leading to problems with digestion and absorption of food; worms don't kill their hosts, they simply make them unthrifty and inefficient. Lungworms are an internal parasite which can cause extremely serious problems. Frequent coughing, thinness, and frothy mucous at the nose indicate this possibility. Stimulate the appetite with warm bran mashes and consult your vet for further treatment. Keep goats on dry pastures; damp, low, or swampy pastures are ideal breeding grounds for parasites. For all parasite problems, consult your vet for dosages and types of wormers to use in your area.

External Parasites

Lice, ticks, screwworms, flies, and maggots can be kept under control, first, by good sanitary procedures—elimination of breeding grounds, such as manure heaps—and by keeping your goats clean and healthy. Signs of external parasites are itchiness, loss of hair, and possibly unthriftiness. Again, there are many dusting powders available for use against these pests; ask your vet about the different ones and their uses.

Foot and Leg Problems

Routine foot care is necessary; goats without access to rocks or hard ground may need their feet trimmed as often as once a month. If your doe becomes accustomed to having her feet handled right from the start, the job should not be too bad, although the hind feet always seem to present more of a problem than the front. In any event, tie her securely and give her something distracting— some hay should do the trick. Softening the hooves beforehand will make it easier, too. (Trim hooves on a rainy day, after the goats have been outside for a while; the hooves will be muddy, but they'll be softened.) Use either a hand pruner or a sharp knife that will lock open. Cutting from heel to toe, trim the bottom of the hoof wall (which may appear turned under) until the hoof is level, trimming just a little at a time. Should you nick a blood vessel in the pad, don't panic: apply pressure to the cut for thirty seconds or so, disinfect it, and proceed. An area becoming pinkish indicates that you've cut away enough of the hoof. Once the outside is even, some trimming may be needed at the back of the hoof; trim it until it is almost pink. The final shape you're aiming for is a trapezoid in cross section, but the final effect you want is a square stance; an untrimmed goat stands awkwardly and may go lame. Again, trimming is something for which you develop a technique, and although you can certainly work it out yourself, the easiest way to learn is to have someone show you the first time around. It's clear sailing from there, especially if you keep up with it so that there is never too much trimming to do at any one time.

If you keep abreast and aware of foot trimming, your goat should develop few foot problems. Foot rot is perhaps the most frequent problem, caused by an organism that thrives in damp conditions. Signs are lameness and soreness on the affected foot, often with a discharge and foul odor as well. To treat it, trim the foot and rinse it with an antiseptic or ointment; sulfathiazole ointment is often used. The best treatment is prevention: keep living quarters clean and dry and keep hooves trimmed. An occasional dusting of the goat's areas with lime will also help.

Surprisingly enough for such agile creatures, goats are subject to sprains and strains in their legs. If your goat seems "off" on a leg, and you've eliminated hoof problems, suspect a sprain or a bruise. If it's not too serious, use liniment on it and wait to see what happens. If it seems to get worse, or if it is severe in the first place, call the vet.

Digestive Problems

Contrary to popular opinion, goats do not have cast-iron stomachs and are, in fact, prone to some digestive upsets. The major ones are scours and bloat, but it's more likely that you'll just come across simple stomachaches. Symptoms and treatment of scours and bloat are the same for goats as for cows. In any case of intense or prolonged scouring, consult a vet immediately. For simple stomachaches in adult goats—which you'll recognize by the goat being less than thrilled at suppertime—a home remedy of Pepto Bismol will probably take care of it. Dose the goat with a bulb baster or similar drenching tool, using the proportional amount per pound as given on the label. Two tablespoons of baking soda dissolved in eight ounces of warm water will also work. Cut out the grain portion of the ration and make sure quality roughage is available until the goat seems to have recovered.

Constipation may occur at times, too. Withhold grain, reduce the roughage, and give a laxative feed such as bran mash or administer milk of magnesia, again using the same dosage per pound as for people. An alternative dose for an adult is two to four ounces of Epsom salts in a pint of warm water; use half as much for a kid. You must administer all drenches slowly, being careful not to hold the animal's head too high. Increase the grain ration gradually once the doe seems normal again.

Other Diseases

Enterotoxemia is a fatal disease caused by a bacterium (*Clostridium perfringens D*) found in the soil and in the digestive tract of nearly every warm-blooded animal. Improper feeding—a big feed of wet grass, too much grain and water, or too much milk—produces conditions conducive to toxemia. The best cure is prevention; vaccination of kids and pregnant does must be part of your routine health program.

Mastitis symptoms and treatment are the same as for the dairy cow.

Milk fever occurs in goats as well as in cows, with much the same causes, signs, and treatment. In goats, however, milk fever may occur as much as a month after kidding; the symptoms will be restlessness and trembling muscles. In the case of milk fever, prop the goat so she'll lie in her normal position (on her midriff) and keep her warm. Call the vet immediately.

Pneumonia in kids is the same as in calves.

Brucellosis and *tuberculosis* are no longer common, but milk goats should be tested yearly for these diseases since they are communicable to humans through milk.

Keep an eye on the general condition of your animals. Problems can range from loneliness (most goats need companionship; rarely can a goat owner own just one goat), to allergies (yes, goats *can* have hay fever), hives and bruises, tooth problems, and the more serious diseases mentioned above. Plant poisoning is a problem to look out for; if your animals are on pasture, particularly browse, be aware of this possibility. Bracken fern, buckwheat, marsh marigold, mountain laurel, buttercup, poison hemlock, rhododendron, azalea, yew, foxglove, delphinium, lily of the valley, and oleander are some of the plants poisonous to goats. It sounds as though the woods are a death trap for grazing goats, but usually, given a choice, goats will naturally avoid these plants. Signs of plant poisoning are vomiting, frothing at the mouth, staggering, convulsions, and pained bleating. Call a vet immediately. In the interim, you can try a purge of four to five tablespoons of Epsom salts in warm water, dosing *very* slowly. If the goat can't swallow, don't force it.

Although in all likelihood troubles will be minor ones, the wise stockkeeper is always aware of the health of her animals.

Milk and Milk Handling

Milking principles are the same for goats as for cows, with some minor variations.

The first variation is the result of height. Since goats are shorter than cows, the quaint little three-legged milking stool becomes relegated to the living room, lowered to the status of a conversation piece. For milking, you have two choices: lower yourself or raise the goat. You *can* milk squatting or sitting on the floor beside the goat, but most people choose to raise the goat instead; enter the milking stand. Easily constructed, the stand has a stanchion to hold the doe and a feedbowl for grain feeding during milking; there is enough room for you to sit next to the doe at a comfortable height on one side; and a leg tie down may be added for a particularly rambunctious goat. (Tie one back leg if the goat is a persistent kicker or fusser during milking.)

Another difference in milking procedure is in the handling of the milk. Although goat's milk is initially cleaner (in terms of bacteria) than cow's milk, it tends to pick up bacteria and odors faster. Therefore, rapid cooling is very important. Some people, after filtering the milk, simply put it in clean glass quart jars in the refrigerator and end up with a perfectly satisfactory product. Others adopt more stringent, though not difficult, cooling techniques. After the filtered milk is placed in clean glass jars, the jar is set in another container into which cold water is run for anywhere from a few minutes to fifteen or twenty. Or the jar can be placed in the larger container, surrounded by ice cubes, and then popped into the refrigerator. Still another method is to put the jars in the freezer

Fig. 4-5. A Stand for Easier and More Sanitary Milking

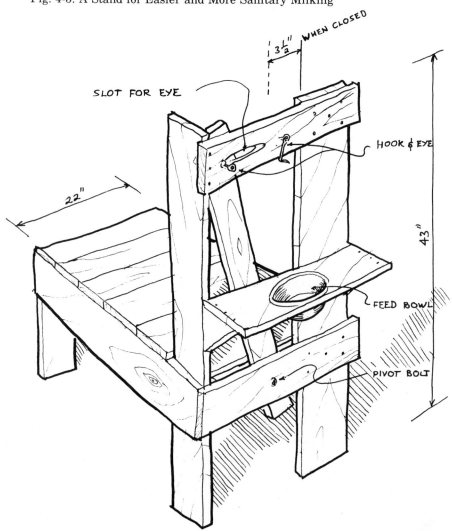

for a while. The common denominator of all these systems, though, is to have the milk in small containers—quarts rather than gallons—to facilitate the rapid cooling.

The "goaty" flavor often attributed to goat's milk is a product of poor sanitation, poor milking methods, and poor milk handling. Properly handled milk from clean, well-kept goats is a rich, sweet-tasting product, often indistinguishable in taste from cow's milk.

A last point concerning goat's milk is the cream. Because of the small size of the fat globules, the cream is not as easily skimmed as cream from cow's milk. So either you get used to using whole milk (not a very difficult task), or if you really want the cream for butter, ice cream, etc., you can use a cream separator. For small quantities of milk, a separator can be a nuisance, but it does the job. If you're milking several goats, the use of a separator is more worthwhile. For small amounts of milk, you can try skimming as the cream rises, but it is a slow process and will take considerable time before you've got enough to do something with other than lighten your coffee.

Any recipes calling for cow's milk can be used with goat's milk, with the exception of cheese recipes. Because of the slightly different nature of the milk, you may need to fiddle a bit with these recipes to make them suitable. Books containing recipes for goat's milk are listed in the references at the end of this chapter.

Meat

When discussing products from the goat, the most obvious is milk, but some consideration must also be given to goat's meat—chevron, as it is technically called.

Goat's meat can be divided into classes by age at slaughter: kid—a few weeks to nine months; mutton—from castrated males nine to eighteen months; billy—from uncastrated males six months and over; and older goat—from older milking does.

Kid is presumably quite good (I've never had any) and is said to be much like lamb. Month-old kid can be treated like veal, three- to four-month-old kid as spring lamb, and six- to nine-month-old kid as lamb or venison. Like venison, older kid needs to be larded, or have fat added to it in some way. Recipes for kid can be found in many French cookbooks.

Goat mutton is similar to ewe mutton or venison and can be used in recipes calling for venison, marinating it in herbs and oil or larding it to add the necessary fat. Older goat's meat can be treated similarly, but because of its toughness, is better used in recipes based on stewing and currying.

For milk and/or meat purposes, a few goats can be a valuable—and entertaining—addition to the homestead.

References

ALTH, MAX. *Making Your Own Cheese and Yogurt.* New York: Funk and Wagnalls, 1973.

BELANGER, J. *Raising Milk Goats the Modern Way.* Charlotte, Vt.: Garden Way Publishing, 1975.

BRANNON, W. F. *The Dairy Goat.* Cornell University Extension Information Bulletin 78. Ithaca: New York State College of Agriculture and Life Sciences.

GREGORY, DIANA. *Dairy Goats.* New York: Arco Publishing Co., Inc., 1976.

MACKENZIE, D. *Goat Husbandry,* 3rd ed. London: Faber and Faber Ltd., 1970.

SALMON, J. *The Goatkeepers Guide.* London: David and Charles Newton Abbot, 1976.

III. RAISING MEAT ANIMALS

5. MEATS

6. BEEF CATTLE

7. SWINE

8. SHEEP

5
MEATS

Some Thoughts on Raising Meat

Raising your own meat can be a rewarding, nutritious, economical, and (most of all) delicious activity. Meat raising and meat eating, however, have come under a lot of criticism lately, on grounds of health, politics, and morals.

The health issue is an example of how fads can sometimes get blown out of proportion. Contrary to what we are frequently led to believe, the linkage of animal fats and cholesterol with heart disease has never been conclusively established. And although animal fats are thought by some people to be involved in heart disease, considered equally, if not more important, are total caloric intake, *all* fat intake, exercise, weight, genetic background, sex, age, and smoking habits. For the average person, avoidance of meat for reasons of health is simply not indicated.

The political question seems to center on the morality of feeding grain to livestock in a world where humans are starving. Clearly, this is a very complex social and political issue which is not easily solved. For many years, cattle have been looked upon as regulators of the U.S. grain supply. When grain is plentiful and cheap, it is fed to livestock; when it is scarce and expensive, naturally, less grain is fed. Merely ceasing to feed the grain to livestock is not going to put the resultant surplus into the bellies of the hungry people of the world. Then, too, there's the point of all the other resources that have no better use than to be converted into meat. The vast rangelands of the West, and all the land that is too steep, rocky, or otherwise too poor for crops, can be put to no better use (at this point) than grazing. The millions of tons of by-products produced annually from the milling of grain for human use, from the brewing and distilling industries, from other food industries, etc. also find an excellent outlet and use as livestock feed. Particularly for the smallholder, this recycling ability of livestock is extremely valuable. What better use of small fields and marginal land is there than for grazing animals? What better use for all that kitchen and garden waste than a few pigs?

For those who feel that it is somehow not right to kill an animal (with the exception, of course, of those whose sincere religious beliefs preclude animal slaughter), consider this: about 50 percent of the time, bull calves are born to dairy cows, ram lambs are born to wool sheep, cockerels are hatched from hens, and billies are born to nannies. What do you do with male animals that aren't kept for breeding? It seems as though raising them for meat is a natural step in the system. Also, if one is going to avoid meat because "I just hate to think of animals dying," one had better avoid *all* animal products, including leather, eggs, wool, cheese, milk, butter, ice cream, yogurt (all dairy products), and even manure for organic gardens, because the rearing of *all* livestock is based on the disposal of unwanted animals. If it weren't, we'd be overrun with livestock—and pretty poor quality livestock, at that.

For the small stock raiser, though, I think the biggest hurdle to overcome in raising your own meat is psychological. I can't begin to count the number of times someone has said to me, "Oh, but how *could* you?"—usually right after saying, "This is the most delicious pork I've ever had! Where did you get it?" How could I? Well, if I

Grazing animals make efficient use of otherwise unusable roughages and land—rangeland, in this case.
Courtesy: American Hereford Association.

couldn't justify doing it myself I'd have a hard time justifying the purchase of nice, clean, unemotional packages from the supermarket, especially when I think of the life I give my meat animals compared to the lives of commercially raised animals. Commercial animals don't lead awful, suffering lives, as is often implied, but neither do they lead the life of Riley. I must confess, I try to give my animals the good life. When I see my pigs, for example, lying in the sun, rooting in the dirt, wallowing in their mud puddle, or frolicking under the water faucet on a hot day, I know they are getting the best life possible, short of being totally free. That knowledge means a lot to me. I *do* shed a tear and feel sad at slaughter time, but I think that I respect, value, and consequently enjoy the meat the more for the tears. Also, without meaning to sound too crass, knowing that it is going to be the most delicious pork ever doesn't exactly hinder the slaughter process, either. And that's how I could.

Terminology in Brief

When dealing with meat animals, there are a few terms you should understand: carcass weights, dressing percent, and cutability, or percent lean cuts.

The first weight dealt with is "on the hoof" or liveweight. After the animal is slaughtered, head, feet, viscera, blood, hide, heart, lungs, and liver are removed. The weight at this point is *carcass weight*. Depending on whether the carcass is weighed immediately or after chilling, it may be hot carcass weight or cold, the cold weight being somewhat less than the hot. *Dressing percent* is the cold carcass weight as a percent of the liveweight just before slaughter.

In beef cattle, dressing percent corresponds to the quality grades, ranging from as low as 40 percent in the cutter and canner grade up to 66 percent in the prime grade. The usual range for young, finished cattle is 58 to 65 percent, with an average of 62 percent. That is, a steer at 1,000 pounds liveweight will have a 620-pound carcass.

In hogs, dressing percent ranges between 70 and 75, with an average of 72 percent; in lambs, 47 to 54 dressing percent corresponds to USDA quality grades good through prime. Thus, a 200-pound hog with a dressing percent of 72 will yield a 144-pound carcass; a 100-pound lamb with a 50 dressing percent will give a 50-pound carcass.

The carcass weight, however, still does not tell you how much meat you'll have on the table. For this figure you need to know the percent lean, or retail cuts, also called *cutability*. Cutability is calculated as the weight of the retail cuts as a percent of the carcass weight, and it depends greatly on the amount of fat trim. In beef cattle, cutability ranges from about 70 to 80 percent. As a percent of the *liveweight,* however, it is about 43 percent on an animal with a 62 dressing percent. A 1,000-pound animal, therefore, will have about 430 pounds of retail cuts—exclusive of the hide, variety meats, blood, edible and inedible fats, "shrink," and bone and meat waste scraps.

In hogs, cutability is around 72 percent. Lean cuts as a percent of liveweight is around 51. From a 220-pound hog, therefore, you'd have approximately 112 pounds of lean cuts.

Lamb has a cutability of about 81 percent; lean cuts as a percent of liveweight is approximately 41. A 100-pound lamb would put about 41 pounds of lean meat in the freezer.

All these figures are especially important to remember when contemplating raising meat animals, either for your own use or for sale, as it is these percentages that indicate approximately how much meat you'll actually end up with.

USDA Grading System*

Although most small-scale stock raisers will not be concerned with the grading system, it is still of interest to know generally how it works. The systems used for grading beef, pork, and lamb are slightly different, but for expediency, I'll just describe the beef system. (The charts later in this chapter give some idea of the other standards.)

Federal grades are of two types: quality and quantity, or yield. The grading of meat is a service provided by the USDA and is entirely optional. Federal inspection, however, which is separate from grading, is not optional and is required for any meat sold. A yield stamp, a grade stamp, and an inspection stamp are placed on the carcass to show that these procedures have been performed.

Quality Grading

The first step in quality grading is to put the carcass in a class: steer, bullock, bull, heifer, or cow. Steer and heifer carcasses can be graded prime, choice, good, standard, commercial, utility, cutter, and canner. Cow carcasses may fall into any of these grades, with the exception of prime. Bullock may be graded as prime, choice, good, standard, or utility.

Quality grades are then based on the maturity of the carcass and its marbling. Maturity is based on bone ossification and the

---

* Much of this material is adapted from the *Cornell Beef Production Reference Manual,* Fact Sheet 9100, Cornell University Cooperative Extension Service.

color and texture of the lean meat. Younger animals still have some cartilage between the vertebrae of the spinal column, and the lean is a fine-textured cherry red. As the animal matures, the meat becomes darker and more coarse textured, and the cartilage hardens into bone. Table 5-1 shows the physiological maturity score correlated to the approximate chronological age.

Table 5-1. Relationship of Physiological Maturity to Chronological Age

PHYSIOLOGICAL MATURITY SCORE	APPROXIMATE CHRONOLOGICAL AGE (MONTHS)
A	9–30
B	30–42
C	42–54
D	54–72
E	72+

The marbling score is based on the amount of intramuscular fat found in a cross-sectional cut of the rib eye between the twelfth and thirteenth rib. The scoring for marbling is like the grading for olives, where you're always wondering which is bigger, jumbo or mammoth. The categories are abundant, moderately abundant, slightly abundant, moderate, modest, small, slight, traces, and practically devoid. The final quality grade is determined by the relationship between the marbling score and the physiological maturity, as shown in Table 5-2.

Table 5-2. Relationship between Marbling, Maturity, and Carcass Quality Grade

	MATURITY				
DEGREES OF MARBLING	A	B	C	D	E
Slightly abundant	prime				
Moderate			commercial		
Modest	choice				
Small					
Slight	good			utility	
Traces					
Practically devoid	standard			cutter	

Maturity increases left to right (A through E).

Yield Grading

Yield grading is rather complicated, but it is supposed to be an objective measure of the cutability of a carcass, that is, what percent of retail cuts the carcass will yield. All classes of animal are graded by the same standard; yield grades range from 1 to 5, with 1 being the highest percent retail cuts and 5 the lowest. Table 5-3 shows the cutability for each yield grade.

Table 5-3. Yield Grade and Cutability

YIELD GRADE	PERCENT RETAIL CUTS*
1	52.3% or more
2	52.3 to 50.0%
3	50.0 to 47.7%
4	47.7 to 45.4%
5	45.4% or less

* Prediction of cutability or yield of boneless, closely trimmed retail cuts from the round, loin, rib, and chuck expressed as a proportion of carcass weight.

Problems in Grading

A major factor in the yield variations is the amount of fat needing to be trimmed from the cuts. Yield grades and quality grades, therefore, are somewhat contradictory. An animal having enough marbling to be graded prime is unlikely to be yield 1. Of graded commercial beef in 1976, most were graded choice, yield grade 3. Since an animal can be fed to any desired quality and yield grade, this figure was the result of cattle feeders trying to mold the most marketable animal. In fact, more beef and leaner beef can easily be put on the market, but because cattle producers are paid according to the grading system, which puts a premium on fat, it is not economical to do so. This problem is the basis of much controversy within the meat industry, especially since the validity of correlating marbling with quality has been severely challenged. Although it has been generally assumed, by producers and consumers alike, that increased marbling goes hand in hand with greater tenderness and juiciness, impressive numbers of studies have indicated that this is not the case. Age, for example, is considered a more valid indicator of tenderness than is marbling.

The marbling-based quality grading system is also one of the big factors in discrimination against beef from dairy animals. Although it may in fact be more tender because of a younger age, a 1,000-pound Holstein cannot grade as well as a 1,000-pound Hereford because it will not have the same degree of marbling. This is one reason why it is uneconomical commercially to raise Holsteins for beef on a large scale, although in Europe, dairy beef is perfectly acceptable.

Pressure is being exerted to change these grading criteria, however. As more and more beef goes into hamburger—and 40 percent of all beef marketed is now marketed as hamburger—a new, leaner beef animal will become more desirable. Consumer pressure, too, is in the direction of leaner beef. Economics, as well as international and domestic social policies, can also affect the type of beef that reaches the market. In terms of feed, it is more expensive to put on fat than lean, so animals fattened to meet the higher grades are more costly. An animal finished on roughages and little or no grain will be a leaner and somewhat older one, producing a carcass that again would be perfectly acceptable on a European market but which would be graded only good here. Although American consumers can learn to accept good grade beef, and in fact are already consuming much of it in the form of ungraded supermarket meat and hamburger, economic forces still dictate that the cattle raisers aim for higher grades. Also, although there is

Slaughter steers U.S. grades, quality and yield.

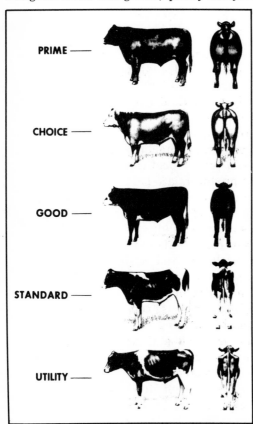
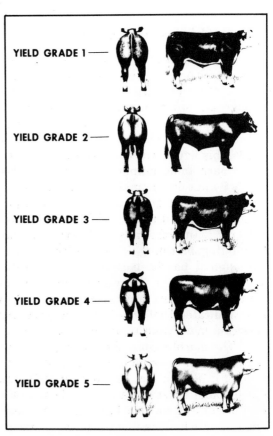

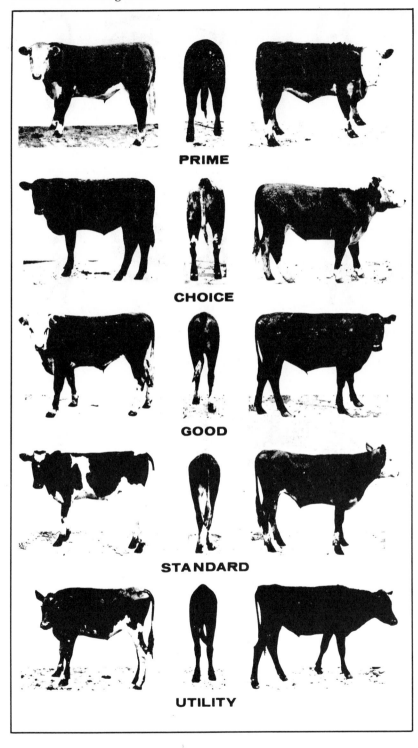

PRIME

CHOICE

GOOD

STANDARD

UTILITY

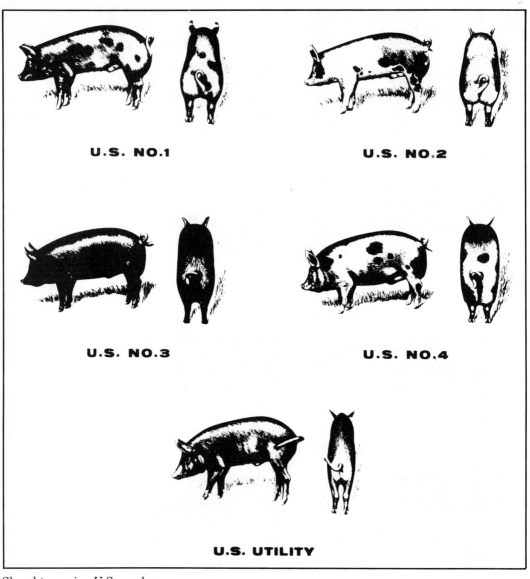

U.S. NO.1

U.S. NO.2

U.S. NO.3

U.S. NO.4

U.S. UTILITY

Slaughter swine U.S. grades.

social agitation against the heavy grain feeding of beef, there are as yet no government policies or economic pressures encouraging a switch to beef finished with roughage. And, of course, as long as the beef producers are paid for fat carcasses, who can blame them for producing such animals?

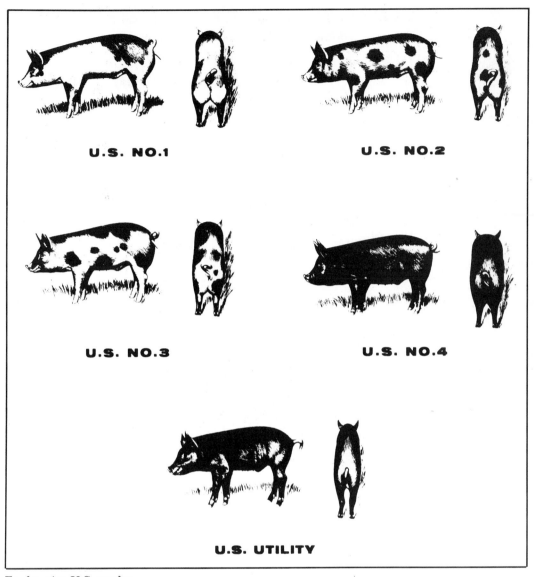

U.S. NO.1

U.S. NO.2

U.S. NO.3

U.S. NO.4

U.S. UTILITY

Feeder pigs U.S. grades.

References

MORTON SALT COMPANY. *A Complete Guide to Home Meat Curing*. Morton-Norwich Products, Inc. Chicago, Ill., 1975.

WELLINGTON, G. H. *Let's Cut Meat*. Cornell University Information Bulletin 92. Ithaca: New York State College of Agriculture and Life Sciences, 1975.

RETAIL CUTS OF BEEF

WHERE THEY COME FROM AND HOW TO COOK THEM

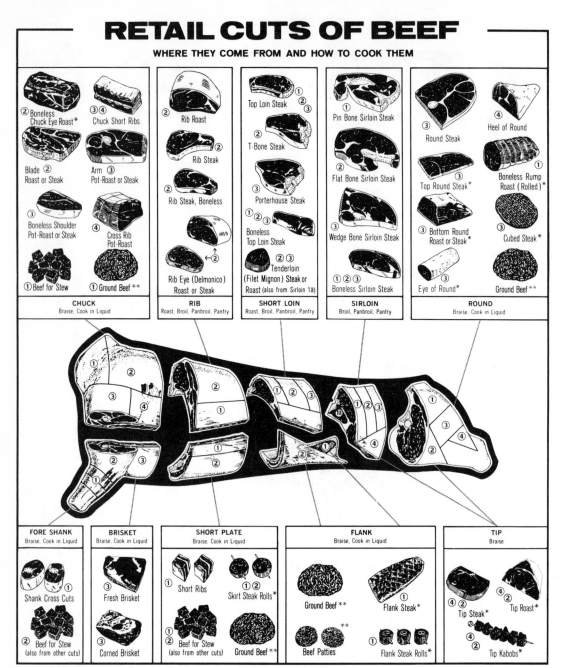

② Boneless Chuck Eye Roast*
③④ Chuck Short Ribs
② Rib Roast
Top Loin Steak ①②③
Pin Bone Sirloin Steak ①
② Round Steak ③
④ Heel of Round

Blade ② Roast or Steak
Arm ③ Pot-Roast or Steak
② Rib Steak
② T-Bone Steak
② Flat Bone Sirloin Steak
② Top Round Steak* ③
① Boneless Rump Roast (Rolled)*

③ Boneless Shoulder Pot-Roast or Steak
② Cross Rib Pot-Roast
② Rib Steak, Boneless
③ Porterhouse Steak
③ Wedge Bone Sirloin Steak
③ Bottom Round Roast or Steak*
② Cubed Steak*

① Beef for Stew
① Ground Beef**
① Rib Eye (Delmonico) Roast or Steak ←②
①②③ Boneless Top Loin Steak
②③ Tenderloin (Filet Mignon) Steak or Roast (also from Sirloin 1a)
①②③ Boneless Sirloin Steak
③ Eye of Round*
① Ground Beef**

CHUCK Braise, Cook in Liquid	**RIB** Roast, Broil, Panbroil, Panfry	**SHORT LOIN** Roast, Broil, Panbroil, Panfry	**SIRLOIN** Broil, Panbroil, Panfry	**ROUND** Braise, Cook in Liquid

FORE SHANK Braise, Cook in Liquid	**BRISKET** Braise, Cook in Liquid	**SHORT PLATE** Braise, Cook in Liquid	**FLANK** Braise, Cook in Liquid	**TIP** Braise

① Shank Cross Cuts
② Beef for Stew (also from other cuts)
③ Fresh Brisket
③ Corned Brisket
① Short Ribs
①② Skirt Steak Rolls*
①② Beef for Stew (also from other cuts)
Ground Beef**
Ground Beef**
① Flank Steak*
Beef Patties**
① Flank Steak Rolls*
④② Tip Steak*
④② Tip Roast*
④② Tip Kabobs*

*May be Roasted, Broiled, Panbroiled or Panfried from high quality beef.
**May be Roasted, (Baked), Broiled, Panbroiled or Panfried.

This chart approved by
National Live Stock and Meat Board

© National Live Stock and Meat Board

Courtesy: National Live Stock and Meat Board, Chicago.

RETAIL CUTS OF PORK
WHERE THEY COME FROM AND HOW TO COOK THEM

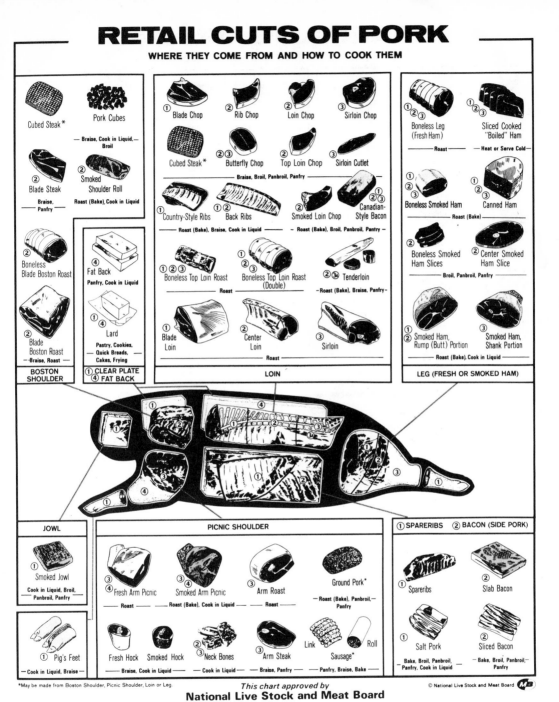

BOSTON SHOULDER

Cubed Steak*

Pork Cubes
— Braise, Cook in Liquid, Broil —

Blade Steak
Braise, Panfry

Smoked Shoulder Roll
Roast (Bake), Cook in Liquid

Boneless Blade Boston Roast

Blade Boston Roast
— Braise, Roast —

① CLEAR PLATE ④ FAT BACK

Fat Back
Panfry, Cook in Liquid

Lard
Pastry, Cookies, Quick Breads, Cakes, Frying

LOIN

Blade Chop

Rib Chop

Loin Chop

Sirloin Chop

Cubed Steak*

Butterfly Chop

Top Loin Chop

Sirloin Cutlet
— Braise, Broil, Panbroil, Panfry —

Country-Style Ribs

Back Ribs
— Roast (Bake), Braise, Cook in Liquid —

Smoked Loin Chop

Canadian-Style Bacon
— Roast (Bake), Broil, Panbroil, Pantry —

Boneless Top Loin Roast

Boneless Top Loin Roast (Double)
— Roast —

Tenderloin
— Roast (Bake), Braise, Panfry —

Blade Loin

Center Loin

Sirloin
— Roast —

LEG (FRESH OR SMOKED HAM)

Boneless Leg (Fresh Ham)
— Roast —

Sliced Cooked "Boiled" Ham
— Heat or Serve Cold —

Boneless Smoked Ham
— Roast (Bake) —

Canned Ham

Boneless Smoked Ham Slices

Center Smoked Ham Slice
— Broil, Panbroil, Panfry —

Smoked Ham, Rump (Butt) Portion

Smoked Ham, Shank Portion
— Roast (Bake), Cook in Liquid —

JOWL

Smoked Jowl
Cook in Liquid, Broil, Panbroil, Panfry

Pig's Feet
— Cook in Liquid, Braise —

PICNIC SHOULDER

Fresh Arm Picnic
— Roast —

Smoked Arm Picnic
— Roast (Bake), Cook in Liquid —

Arm Roast
— Roast —

Ground Pork*
— Roast (Bake), Panbroil, Panfry —

Fresh Hock

Smoked Hock
— Braise, Cook in Liquid —

Neck Bones
— Cook in Liquid —

Arm Steak
— Braise, Panfry —

Link   Roll

Sausage*
— Panfry, Braise, Bake —

① SPARERIBS ② BACON (SIDE PORK)

Spareribs

Slab Bacon

Salt Pork
Bake, Broil, Panbroil, Panfry, Cook in Liquid

Sliced Bacon
— Bake, Broil, Panbroil, Panfry —

*May be made from Boston Shoulder, Picnic Shoulder, Loin or Leg.

This chart approved by
National Live Stock and Meat Board

© National Live Stock and Meat Board

Courtesy: National Live Stock and Meat Board, Chicago.

123

RETAIL CUTS OF LAMB

WHERE THEY COME FROM AND HOW TO COOK THEM

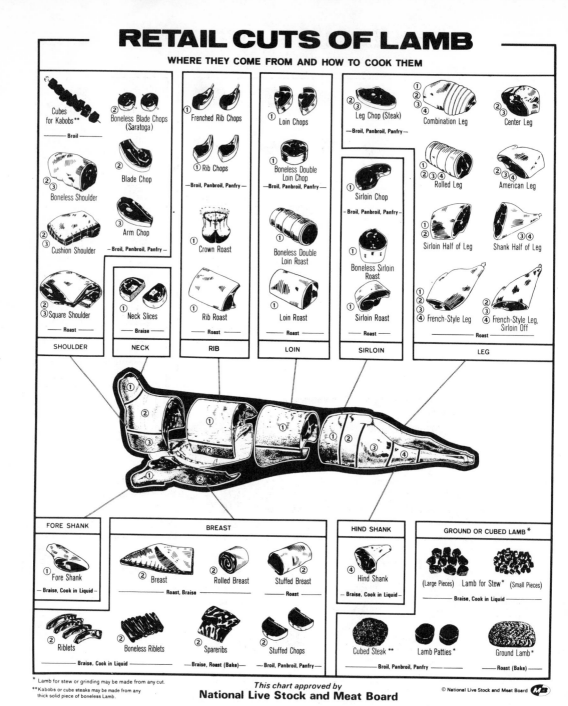

SHOULDER

Cubes for Kabobs**

② Boneless Blade Chops (Saratoga)

— Broil —

②③ Boneless Shoulder

②③ Cushion Shoulder

②③ Square Shoulder

② Blade Chop

③ Arm Chop

— Broil, Panbroil, Panfry —

— Roast —

NECK

① Neck Slices

— Braise —

RIB

① Frenched Rib Chops

① Rib Chops

— Broil, Panbroil, Panfry —

① Crown Roast

① Rib Roast

— Roast —

LOIN

① Loin Chops

① Boneless Double Loin Chop

— Broil, Panbroil, Panfry —

① Boneless Double Loin Roast

① Loin Roast

— Roast —

SIRLOIN

③ Leg Chop (Steak)

— Broil, Panbroil, Panfry —

① Sirloin Chop

— Broil, Panbroil, Panfry —

① Boneless Sirloin Roast

① Sirloin Roast

— Roast —

LEG

②③④ Combination Leg

②③ Center Leg

①②③④ Rolled Leg

②③④ American Leg

①② Sirloin Half of Leg

③④ Shank Half of Leg

①②③ French-Style Leg

②③④ French-Style Leg, Sirloin Off

— Roast —

FORE SHANK

① Fore Shank

— Braise, Cook in Liquid —

② Riblets

— Braise, Cook in Liquid —

BREAST

② Breast

② Rolled Breast

② Stuffed Breast

— Roast, Braise —

— Roast —

② Boneless Riblets

② Spareribs

② Stuffed Chops

— Braise, Cook in Liquid —

— Braise, Roast (Bake) —

— Broil, Panbroil, Panfry —

HIND SHANK

④ Hind Shank

— Braise, Cook in Liquid —

GROUND OR CUBED LAMB *

(Large Pieces) Lamb for Stew* (Small Pieces)

— Braise, Cook in Liquid —

Cubed Steak **

Lamb Patties *

Ground Lamb *

— Broil, Panbroil, Panfry —

— Roast (Bake) —

* Lamb for stew or grinding may be made from any cut.
**Kabobs or cube steaks may be made from any thick solid piece of boneless Lamb.

This chart approved by
National Live Stock and Meat Board

© National Live Stock and Meat Board

Courtesy: National Live Stock and Meat Board, Chicago.

124

RETAIL CUTS OF VEAL

WHERE THEY COME FROM AND HOW TO COOK THEM

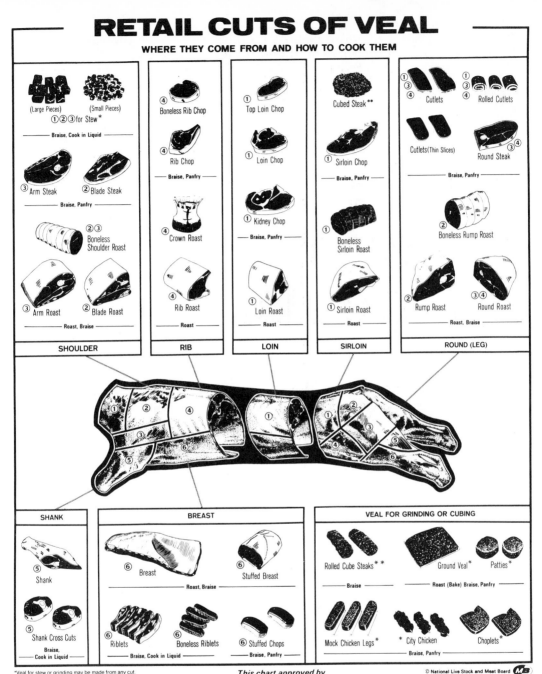

(Large Pieces) (Small Pieces)
①②③ for Stew*
— Braise, Cook in Liquid —

③ Arm Steak   ② Blade Steak
— Braise, Panfry —

②③ Boneless Shoulder Roast

③ Arm Roast   ② Blade Roast
— Roast, Braise —

SHOULDER

④ Boneless Rib Chop

④ Rib Chop
— Braise, Panfry —

④ Crown Roast

④ Rib Roast
— Roast —

RIB

① Top Loin Chop

① Loin Chop

① Kidney Chop
— Braise, Panfry —

① Loin Roast
— Roast —

LOIN

Cubed Steak **

① Sirloin Chop
— Braise, Panfry —

① Boneless Sirloin Roast

① Sirloin Roast
— Roast —

SIRLOIN

① Cutlets   ① Rolled Cutlets

Cutlets (Thin Slices)   Round Steak
— Braise, Panfry —

② Boneless Rump Roast

② Rump Roast   ③④ Round Roast
— Roast, Braise —

ROUND (LEG)

SHANK

⑤ Shank

⑤ Shank Cross Cuts
Braise, Cook in Liquid

BREAST

⑥ Breast   ⑥ Stuffed Breast
— Roast, Braise —

⑥ Riblets   ⑥ Boneless Riblets   ⑥ Stuffed Chops
— Braise, Cook in Liquid —   — Braise, Panfry —

VEAL FOR GRINDING OR CUBING

Rolled Cube Steaks **   Ground Veal*   Patties*
— Braise —   — Roast (Bake) Braise, Panfry —

Mock Chicken Legs*   * City Chicken   Choplets*
— Braise, Panfry —

---

*Veal for stew or grinding may be made from any cut.

**Cube steaks may be made from any thick solid piece of boneless veal.

This chart approved by
National Live Stock and Meat Board

© National Live Stock and Meat Board

Courtesy: National Live Stock and Meat Board, Chicago.

125

6
BEEF
CATTLE*

You go to the supermarket and pick up a steak. It may be graded or not, but it looks good enough. Maybe you'll look at the price and put it down again; maybe you'll take it home and eat it; maybe you'll think it was a good steak, or maybe you won't. Where did that steak come from? What kind of animal, from what conditions, and fed on what? Buying a steak can be chancy every time, because often you have no information about the quality of that steak other than the price and the looks. This is one big advantage to raising your own beef: you know what the animal has eaten, you control the cuts, and you can at least understand, if not keep down, the price.

The beef industry is extremely complex. We get our steaks and McDonalds gets its patties through different and complicated pathways. However, by and large the most common source of our beef is the feedlot, where weaned calves, bought from cow-calf operations, are fattened on grain until they reach the optimum slaughter weight of about one thousand pounds. The number of animals coming through the feedlots as opposed to the other routes (Fig. 6-1) is greatly influenced by the price of grain: when grain is cheap and plentiful, more animals go through the feedlots; when it

* Much of the information in this chapter is based on and adapted from extension publications of the New York State College of Agriculture and Life Sciences, Cornell University, with thanks to Mark Semlek and Michael Thonney.

is scarce and expensive, the other pathways get heavier use. Generally, in this country, we have a surplus of grain, which allows the feedlot system to operate. In other countries, however, and at times in this country, grain is scarce, and human consumption of the grain has priority; grass finishing then becomes the common method of beef production, as opposed to grain finishing in the feedlot.

Cow-Calf to Feedlot—The Beef Production System

The purpose of the cow-calf operation is to provide weaned feeder calves to another type of producer, the feedlot operator, to finish as slaughter cattle for the beef trade. The cow-calf operator owns a beef herd, which is maintained, normally, on range and pasture. Most range is in the West, and consequently most cow-calf opera-

Fig. 6-1. The Beef Route

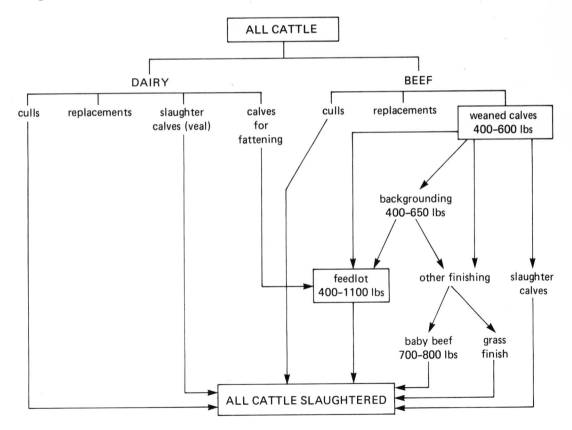

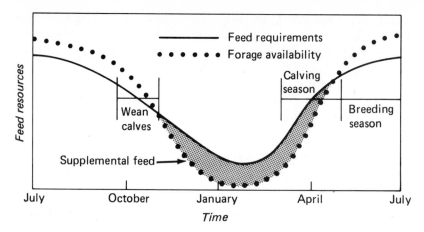

Fig. 6-2. Reproductive Year in Relation to Nutrient Needs and
Availability of Forage
Source: M. A. Semlek, *Beef Production,* Extension Publication 133, N.Y.S.
College of Agriculture and Life Sciences, Cornell University.

tions are located in the West—often extensive operations involving
thousands of acres. With the longer grazing season found in the
South and the possibilities of improved pastures, more intensive
cow-calf operations are recently being established there as well.

The cow herd is maintained through the winter with minimal
shelter and on a diet consisting primarily of roughage, with some
protein supplement as necessary. Cows calve in the spring, and the
calves remain with their dams until fall. Fall is weaning time,
when calves are six to eight months old and weigh 400 to 650
pounds; these are the "feeder," or "weaner," calves. The spring-
calving, fall-weaning schedule evolved so that the cows' nutritional
needs would correspond to the forage supply (Fig. 6-2). When nutritional demands are highest (lactation and rebreeding), the forage
supply is at its best (spring pastures); when nutritional needs are
lowest (early gestation), the cows are fed harvested roughages (winter feeding). This system works out most economically for the cattle
producer.

After weaning, the calves can follow any of various pathways.
Usually, they are bought by a feedlot and fed primarily grain rations for another six to seven months, after which they are sold to
slaughterhouses and processors as finished cattle at twelve to fifteen
months old. Sometimes the cow-calf operator, or perhaps someone
else altogether, will have a surplus of roughages available for the
winter. In that situation, weaned calves may be wintered on the
roughages, then either sold to the feedlot as yearling cattle to be
fed to the proper finish, or finished on grain, summer pasture, or
pasture plus grain and then sold directly by the owner.

Home Beef Production

The person thinking of raising beef on a small scale can find a place in this system as the owner of a small cow-calf herd—finishing the beef as well (being cow-calf and feedlot operation in one)—or as a miniature feedlot operator—buying a weaned calf and finishing it. Other options, which do not fit into the traditional beef-rearing patterns, are also possible and are described later. The system you follow obviously depends on your resources. A cow-calf unit needs approximately three acres of decent pasture, and the cow will consume about 1½ tons of hay over the winter. Grain, machinery, buildings, veterinary services, supplies, equipment, labor, and land costs all must be considered. The person with limited land resources who must purchase all forage is not really in a position to keep brood cows, but should instead consider buying a weaned feeder animal to raise to supply the family's beef.

Whatever system you choose, however, do not expect to make carloads of money at it. Maintaining a small herd or using one of the nontraditional methods of beef raising might net you a small income, but basically the person raising beef at home should be doing so for the better quality product and the potential lower costs (which are only *potential*; homegrown beef can range from half to twice the price of supermarket beef, depending on the resources and the skills of the manager). Selling some beef, usually through the freezer trade, can offset the cost of raising your own.

Hereford cattle.
Courtesy: American Hereford Association.

Breeds

Beef cattle fall into four major divisions: European, Indian, U.S. developed, and exotic breeds. Some breeds within the exotic and European classes are considered dual-purpose milk and meat animals. Altogether, there are about forty major beef breeds, but new ones are constantly being developed through crossbreeding. The most common breeds are mentioned below.

European breeds include the Angus, Hereford, Shorthorn, Milking Shorthorn, Red Angus, Red Poll, and Devon. Angus are small (cows weigh about 1,500 pounds), black, polled (hornless) cattle which originated in Scotland. They are supposed to be good mothers and milkers, cold tolerant, easy calvers, and good foragers; they also produce an excellent carcass with little fat, small bones, and heavy muscling. The small size, however, is seen as a drawback by many cattlemen.

The white-faced cattle you've probably seen around are Herefords: white face, throat, brisket, flank, switch, and lower legs with a red body. Originating in England, the Hereford is larger than the Angus (cows weigh 1,700 pounds) and is considered to be hardy and adaptable with good grazing ability. Low milk production and eye problems (caused by lack of pigment around the eyes) are the major criticisms leveled against Herefords.

Shorthorns are large cattle (cows at 2,000 pounds) with red, white, or roan coats. The Shorthorn also came from England, is a good milk producer, gains well in the feedlot, and utilizes roughages efficiently, but it can tend to produce a somewhat inferior carcass if carried too long. The Milking Shorthorn is similar to the Shorthorn but is, obviously, a dual-purpose animal.

Santa Gertrudis bull.
Courtesy: Santa Gertrudis Breeders' International.

Brahman are not merely a different breed of beef cattle but a different species altogether—*Bos indicus* as opposed to *Bos taurus*— and as such, are widely used for crossbreeding. The Brahman, although originating from the Indian breeds, was developed in the United States. Colors range from light gray to dark, and the body type is characterized by a hump on the neck, long ears, and a large, loose dewlap. The strong points of Brahman cattle are many: heat tolerance, resistance to disease and parasites, ability to twitch their skin (which is a boon in insect-plagued areas), and good mothering ability. Poor fertility is a problem in some bloodlines, and poor cold tolerance is considered another weak point.

Breeds developed in the United States include the Santa Gertrudis, Brangus, Red Brangus, Beefmaster, Braford, Charbray, Barzona, Polled Shorthorn, and Polled Hereford. The Santa Gertrudis is a large (1,600 pounds) red animal with a loose hide similar to that of a Brahman. Not surprisingly, it was developed as a cross between the Shorthorn (five-eighths) and the Brahman (three-eighths). Good beef production on grass, excellent foraging ability, heat tolerance, resistance to insects and disease, and good beef conformation are the virtues of this breed. Brangus are (take a guess) three-eighths Brahman and five-eighths Angus: black, polled cows with the Brahman hump. The Beefmaster is another new breed utilizing Brahman blood (one-quarter Hereford, one-quarter Shorthorn, one-half Brahman), as is the Braford (five-eighths Hereford, three-eighths Brahman). Polled Herefords are quite similar to Herefords minus the horns, as the Polled Shorthorns are similar to Shorthorns without horns.

Beef breeds considered exotic include animals from France (Charolais, Blonde D'Aquitaine, Limousin, Maine Anjou), Switzer-

Brahman cow.
Courtesy: American Brahman Breeders Association.

Brahman hybrid steer.
Courtesy: American Brahman Breeders Association.

Shorthorn bull.
Courtesy: American Shorthorn Association.

Milking Shorthorn cow.
Courtesy: American Milking Shorthorn Association.

Angus bull.
Courtesy: American Angus Association.

Brangus cow and calf.
Courtesy: International Brangus Breeders Association, Inc.

land (Brown Swiss, Simmental), England and Scotland (Galloway, Lincoln Red, Scotch Highland, Sussex, Welch Black), Italy (Chianina, Marchigiana, Piedmont), Austria (Gelbviech, Pinzgauer), Australia (Murray Grey), and Canada (Hays Converter). Many of these breeds are used to meet specific conditions (such as the use in North Dakota of the Galloway, a hairy, hearty breed well suited to the cold); some are dual-purpose animals—or even triple purpose, for milk, meat, and draft—(Brown Swiss, Chianina, Gelbviech, sometimes Simmental), but most are used in crossbreeding.

As with swine and sheep, beef producers may maintain purebred or commercial herds. The purebred, registered lines are kept essentially as a source of superior breeding stock. Commercial beef producers, however, generally use grade or crossbred animals, following any of a variety of breeding programs, often involving purebred bulls on crossbred cows. Crossbreeding is becoming more widely used in the beef industry because of the significant effects of heterosis (see Chapter 2), particularly on the percentage of calves weaned and weaning weights. Straightbred, nonregistered cattle, however, are still extremely popular among commercial cattlemen.

As always, the breed of animal used is the personal decision of the owner, based on breed characteristics, suitability to the enterprise, availability, economics, and to a great extent, personal preference. Although any beef producer could find a suitable animal among the beef breeds, some mention must be made of the possibility of using dairy animals for beef—particularly for the small producer. Contrary to popular opinion, dairy steers (especially Holsteins) can make quite as good beef as can the beef breeds. Although Holsteins were once criticized for their muscle distribution, it has since been shown that, in fact, the muscle distribution of a Holstein is the same as in a beef breed. Because most of the expensive cuts come from the hindquarter, it is important to note that both Holsteins and beef breeds have 53 percent of their muscle in the hindquarter. An advantage of Holsteins is that they grow faster; they will reach 1,000 pounds before a beef breed animal. However, a 1,000-pound Holstein will have less fat than a 1,000-pound Hereford (for example) and will not be graded as well; the beef animal might be graded choice, whereas the Holstein would be graded good or low choice. In order to get more fat on the Holstein, it would have to be carried to a heavier weight; perhaps a 1,300-pound Holstein would be graded like a 1,000-pound Hereford. Also, with less of a fat cover, the Holstein carcass could not be aged as long as a carcass from a fatter animal—a distinct disadvantage, particularly for the restaurant trade.

Because of the fat issue, and the consequent low grading of dairy beef, raising them on a large scale is uneconomical. For the small-scale producer, however, who is producing for home use and limited direct sales to consumers, and therefore is not involved with the grading system, dairy beef might be quite economical. In fact, many people actually prefer the leaner carcass; a 1,000-pound lean carcass will also yield better than a 1,000-pound fat carcass. Dairy beef, therefore, definitely should not be ruled out by the small-scale producer (especially since marbling, or intramuscular fat, has never been well correlated to quality in the first place).

Management of a Brood Cow Herd

Shelter and Equipment

Beef cattle are extremely well adapted to adverse conditions and so have minimal housing needs. During the summer, shade must be available, and during the winter, anything from a sheltered woodlot to a three-sided shed to an enclosed barn will be suitable. Where winter conditions are severe, cattle will do better if offered more, rather than less, protection, but also avoid going overboard: beef cattle grow a heavy winter coat and will not do well in too tightly confined, poorly ventilated barns. Dairy or dairy crossbred cows will need somewhat more protection. Allow each animal, when confined, at least seventy-five square feet of space.

Feeding and watering equipment can also be simple: hay can be fed in racks, grain in buckets or bunkers, and water from tubs, troughs, or buckets. Hayracks designed with a shelf beneath will help reduce waste, especially when feeding leafy dry hay, where leaves tend to shatter and get trampled underfoot. The shelf will catch the leaves (the most nutritious part of the hay), and the cows will lick them off. Grain feeders with keyhole openings can help assure each cow of her fair share of the grain. Waterers must

Beef cattle are hardy, but some simple shelter for your animals
during severe weather is of value.
Photo: American Hereford Association.

provide ample clean water at all times. Salt blocks and mineral
boxes (when necessary) should be provided.

Some equipment for handling the cows and calves will be
needed. Because beef cattle are handled much less than dairy cattle,
they are often less tractable; with only a few cows, however, handling should not be much of a problem. With more than just a
couple of animals, some type of permanent or portable corral and
squeeze chute is necessary. The cows are herded into an area where
the only exit is an alley, the width of a single cow, which leads
either to another pen or to a stanchion—or to any other way to
restrain the animal as an individual. The setup will be as simple
or as complex as necessary to handle the number and type of animals you have, but *some* well-designed, easy-to-use method of gathering, holding, and restraining individual cows and calves must be
available.

Equipment for dehorning, castrating, identifying (if necessary), dosing, and treating cows and calves is also necessary (specific
items needed are covered later in this chapter).

Breeding

The annual reproductive cycle in the beef cow can be divided
into three periods: gestation, calving, and rebreeding. The time of
year that these phases occur depends on the calving season. Spring
calving is the norm in the cattle business, but it is by no means

Table 6-1. Spring and Fall Calving Schedules

	GESTATION	LATE GESTATION	CALVING–REBREEDING	END OF BREEDING TO WEANING
Spring calving	Nov.–Mar.	Feb.–Mar.	Apr.–Aug.	Sept.–Nov.
Fall calving	May–Sept.	Aug.–Sept.	Oct.–Feb.	Mar.–May

sacred; fall calving programs are sometimes followed (Table 6-1).

The idea behind spring calving is that the cows' peak nutritional needs (calving through rebreeding) occur while the pastures are at their best and the lowest demands (gestation) correspond to the period when forages are poorest. Having cows calve on pasture also means less shelter is needed, and frequently disease and nutritional problems are also reduced. Fall calving, however, allows for marketing weaned calves in the spring, when feeder cattle prices are highest, or allows the alternative of pasturing the calves all summer to sell in the fall as yearlings, if that procedure is more advantageous. Also, fall calving does not conflict with heavy spring planting schedules. Of course, the big drawback to fall calving is the additional management needed for the cows; the peak nutritional needs must be met with harvested feeds, so much more attention is needed to ensure the quality of the rations.

Most of the breeding information in Chapter 3 applies to beef cattle as well. The estrous cycle is approximately twenty-one days, with the heat period (the time the cow can be bred) lasting sixteen to twenty hours. Signs of heat are cows mounting other cows or standing to be mounted (of two cows, the one allowing herself to be mounted is the one in heat), swelling of the vulva, mucous discharge, and frequent urination. The big difference, though, is that in beef herds, usually a bull takes care of estrus detection, not a human; artificial insemination, although quite common in dairy herds, is not widely used in beef cattle because of the additional management skills it requires. In order to have a successful A.I. program, the cows must cycle and show heat regularly, and the heat periods must be detected by the manager. With proper nutritional levels, cows should show heat three to four weeks after calving, but catching them in heat is still tricky. Heat mount detectors can be used in conjunction with morning and evening observation of the animals. (Heat mount detectors are dye-filled tags, which are attached to the animals' tailhead; when the cow is mounted, the dye is released.)

Cows should be bred twelve hours after being seen in heat; clearly, some method of restraining the cows for breeding must be available. With small herds, very good conception rates can be achieved using A.I. if estrus detection is good; but with larger herds

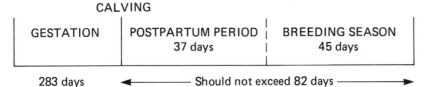

| CALVING | | |
| GESTATION | POSTPARTUM PERIOD
37 days | BREEDING SEASON
45 days |

283 days ←——————— Should not exceed 82 days ———————→

Fig. 6-3. The Ideal Reproductive Year

or poorer estrus detection, "clean-up" bulls might be needed to naturally breed the unbred cows. For the person with just a few brood cows, extra vigilence in heat detection, enabling the full use of A.I., is well worth the trouble compared with the cost and trouble of keeping a bull. At the very least, the owner of a small herd should make every effort to rent a bull during the breeding season rather than keeping one all year just to breed a few cows.

When using natural service, a proper cow-to-bull ratio must be maintained. A well-grown, year-old bull can breed ten to fifteen cows in a sixty-day breeding season, and a mature bull in good condition can breed up to forty. The usual ratio, however, is one bull for twenty-five cows.

Another difference in the breeding of beef versus dairy cows is that in a beef herd, you want all the cows to calve as close to the same time as possible which is achieved by the use of a restricted breeding season. This system allows for greater labor efficiency, shorter calving intervals, and more uniform calves at weaning time. A forty-five- to sixty-day breeding season—with the aim being to get as many cows bred back as early in the season as possible—will keep the calving interval at one year or less. Forty-five to sixty days gives the cow three heat periods in which to be bred; if a cow doesn't settle within that time, she should be considered a prime candidate for culling. (The ideal situation is shown in Fig. 6-3.)

After the cows have been bred, they should be checked to make sure they have settled. Pregnancy testing by a vet is reliable and usually inexpensive.

It has already been stated that cows are bred starting about forty days after calving. What about heifers? At breeding time, heifers should weigh a minimum of 650 to 700 pounds, which they will when they're twelve to fifteen months old if they've been well managed. Because first-calf heifers often have more calving difficulties than do older cows, they are bred to calve a few weeks before the cow herd; the manager can then put in extra time as necessary with the heifers. Also, since first-calf heifers tend to have smaller calves as well, the extra few weeks headstart for those calves results in more uniformly sized calves at weaning time. A third reason is that earlier calving gives the heifers extra time postpartum to get into condition for rebreeding.

Calving and Calf Management

Calving time is exciting, be it your one dairy cow giving birth to her one annual calf or your fifteen beef cows all calving within a short period of time. To me, the sight of healthy, vigorous young ones—of whatever species—is beautiful, and a field of contentedly grazing beef cows, with little calf faces poking through the grass here and there, is right up there.

Management of the beef cow at calving time is similar to that of a dairy cow. The basics are: observe the cow (and use your records, if you've bred with A.I.) for signs of approaching calving; have cows calve in clean, dry, well-ventilated but draft-free calving quarters, or ideally, on clean pasture on a nice day; be aware of potential calving problems and know how to deal with them; wipe mucous from the nose of the newborn and dip the navel in iodine; and see that the calf nurses within a few hours. Weigh the calf within a few days, if possible, and keep accurate records. Later practices to be carried out on the calves include castration and dehorning (see Chapter 3 for dehorning methods).

Castration of male calves should be done by the time they are eight to ten weeks old; the younger the calf, the less stress will occur. Early spring or late fall is the best time for castration in terms of disease and flies. If a calf must be castrated when flies will be troublesome, one of the bloodless techniques is recommended. If this is not possible and there is an open wound on the calf during fly season, apply fly repellent and watch the calf closely until the wound is healed. The position in which to hold the calf and the methods of castration are outlined in a reprint from the Cornell Beef Production Reference Manual:

> Castrating is removing the testicles of male calves. Male calves in commercial herds should be castrated before they are 8 to 10 weeks of age. It is less stressful to the calf and less work for the producer if the calves are castrated when young. However, in purebred herds, it is desirable to let bull prospects grow to 6 to 8 months of age before making a final decision on which calves should be castrated. If a bull calf is not castrated before he is 10 to 12 months old, he may become "staggy," which is considered an undesirable characteristic in a feeder or finished steer sold through traditional commercial markets.
>
> At the present time, producers castrate calves at slightly older ages to take advantage of the higher gains and greater efficiency of bulls and also to lessen the hazard of urinary calculi.[1]
>
> *Season to castrate:* Castration should be done in early spring or late fall to avoid infestation of flies. If screwworms

---

[1] Ensminger, M. E., *Beef Cattle Science,* 4th Ed., 1968. The Interstate Printers and Publishers, Inc., Danville, IL.

are prevalent in your area, fly repellent should be applied to the wound and the animal should be kept under close supervision until the wound has healed. A bloodless method of castration also can be used to prevent infection.

Positions to restrain calves when castrating: Young calves are usually thrown to be castrated. Animals of 6 months of age or older can be more easily operated on in a standing position with their heads secured in a head gate or stanchion.

Calves that are in large pastures should be corralled and separated from the main herd or the calves can be roped. Roping should be done quietly by an expert without disturbing the rest of the herd.

After the animal is separated and thrown, two people should hold down the animal: one person kneeling on the calf's shoulder or neck and holding the top front leg; the other person sitting at the rear of the calf pushing his foot against the calf's leg closest to the ground and pulling the other leg back toward himself. A third person then leans over the back of the animal and proceeds with the castration.

METHODS OF CASTRATING

1. *Knife.* Most people prefer to use a sharp knife to castrate their calves. This method assures that the animal is completely castrated and permits ample drainage of the affected area.

The operator should cleanse his hands with an antiseptic solution and the castrating instruments should be kept in solution. The scrotum of the animal should also be washed with an antiseptic solution before castrating.

a. *Removal of the lower end of scrotum.*
1. Grasp the scrotum with one hand and cut off the lower third of the scrotum with a sharp knife.
2. Push the testicle out of the scrotum.
3. The membranes covering each testicle are slit or these membranes can simply be removed along with the testicle.
4. In younger calves, the testicles can be pulled from the body.
5. In older calves, the cord is severed with a scraping motion using a knife.
6. Alternatively, steps 3 through 5 can be accomplished with an instrument known as an emasculator. This instrument has a cutting edge and crushing edge. The crushing edge is placed toward the animal (away from the testicle). If properly performed, the emasculator method should reduce the amount of bleeding as a result of the castration.

7. Dust the wound with sulfa-urea powder to aid in healing.
8. If flies are bothersome, a creosol compound can be swabbed on the wound or wound spray can be used.

b. *Slitting the scrotum down the sides.*
  1. One testicle is pulled down at a time and is held firmly to the outside so that the skin of the scrotum is pulled tight over the testicle.
  2. Using a sharp knife, slit the outside of the scrotum next to the leg.
  3. Make sure the end of the scrotum has been cut to allow for proper drainage.
  4. Remove the testicle using the same scraping motion as described above or use an emasculator.
  5. Repeat on the other side and treat with an antiseptic solution.

2. *Elastrator.*
a. Push both testicles into the lower end of scrotum.
b. Using the elastrator, press both testicles through the rubber ring.
c. Release the rubber ring 1½ to 2 inches above the testicles.
d. The testicles and scrotum will slough off within 1½ to 2 weeks.
e. This method has been criticized because it is not as sure as the surgical method and because an open wound may result when the tissue is sloughed off.

3. *Burdizzo pincers.* This is a bloodless method of castration. The cords and associated blood vessels are crushed or severed so that the testicles degenerate from lack of circulation.

a. The cord of one testicle is worked to the side of the scrotum.
b. Clamp the Burdizzo 1½ to 2 inches above the testicle and hold for a few seconds.
c. Repeat this operation on the same testicle. Move the Burdizzo ¼ inch lower and clamp again.
d. Repeat this procedure on the other side.

It is important to clamp only the cord so the circulation to the central part of the scrotum will not be affected. If the cord is incompletely crushed, the animal may develop "stagginess" later on. Since this is a bloodless method, there is no trouble with flies or screwworms.*

---

* Amy I. Anderson, "Castration of Cattle," *Cornell Beef Production Reference Manual,* Cornell University Fact Sheet 5510, May 1977.

Record Keeping

In a herd of any size, accurate records should be kept to facilitate intelligent management decisions, particularly about breeding and culling. Depending on the size of your herd, some form of identification of the cows may be necessary (by name is fine for the small herd). The identification system can be anything from numbered ear tags or neck chains to branding or ear tattoos. Once you can identify each cow in your herd, the important information to be recorded is as follows:

1. The cows' reproductive efficiency, that is, the calving interval, or the amount of time between consecutive calvings.
2. Calving data: exact birth date of the calf, weight, sex, sire, and calving problems.
3. Weaning weight: weigh calves at weaning, approximately 205 days (record the date weighed); this figure indicates the milk-producing ability of the cow and allows for the basic calculation of how much feed it took to produce a pound of weaned calf.
4. Average daily gain should be calculated as follows:

$$ADG = \frac{weaning\ weight - birth\ weight}{age\ in\ days}$$

   This equation represents the growth rate of the calf.
5. If calves are kept through a postweaning period, yearling weights (and the date of the weighing) should be recorded. Postweaning average daily gain is calculated as follows:

$$PWADG = \frac{yearling\ weight - weaning\ weight}{number\ of\ days\ between\ weighings}$$

Feeding

As with all livestock, the level of feeding for the beef cow is based on her productive state. During gestation, only maintenance requirements must be met, as the demands of the fetus are slight; roughages can supply most of the necessary nutrients. Two months before calving, nutritional needs increase, since this is the period of rapid fetal growth; additional grain supplementation may be needed to get the cow in good (but not fat) condition for calving. Calving through rebreeding represents the time of greatest nutritional demands; lactation demands are high, plus the cow must be in a gaining condition for optimal rebreeding. Once the cow is bred, the critical feeding phase has passed. From rebreeding through weaning, lactation needs must still be met, but by this time, calves are consuming other feeds in addition to mother's milk. The next feeding period brings the cycle back around to the dry, late gestation cow.

Replacement heifers, fed to gain one to one and one-half pounds per day from weaning at seven to eight months to breeding at twelve to fifteen months (or 650 to 700 pounds), should be fed separately from the cow herd. Table 6-2 gives the nutrient requirements of beef cattle during the different stages of production. This table can be used to calculate rations as shown in Appendix 1, or the suggested guideline rations can be used. The rations are just that, however: guidelines. It goes without saying that the manager, based on daily appraisal of the cattle, must adjust these rations as necessary.

Table 6-2. Nutrient Requirements of Beef Cattle

DAILY NUTRITIONAL REQUIREMENTS

Body Wt. (LBS.)	Minimum Dry Matter Intake (LBS.)	Crude Protein (LBS.)	TDN (LBS.)	Calcium (GRAMS)	Phosphorus (GRAMS)	Vit. A (THOUSANDS I.U.)
Pregnant mature cows (middle ⅓ of pregnancy)						
900	14	.80	7.3	11	11	17
1,000	15	.86	7.8	12	12	19
1,100	16	.92	8.3	13	13	20
1,200	17	1.00	8.8	14	14	22
Dry, pregnant mature cows (last ⅓ of pregnancy)						
900	16	1.0	8.3	14	14	21
1,000	17	1.1	8.8	15	15	23
1,100	18	1.1	9.4	15	15	24
1,200	19	1.2	9.9	16	16	26
Lactating cows (first 3–4 months after calving)						
900	22	2.2	11.8	35	33	28
1,000	23	2.3	12.3	36	34	30
1,100	24	2.4	12.8	37	35	31
1,200	25	2.6	13.4	37	36	34
Weaned heifer calves (average daily gain of 1.5 lbs.)						
330	9	1.1	6.2	18	14	9
440	13	1.3	8.4	18	16	13
550	13	1.4	9.1	17	15	14
660	15	1.5	10.4	16	15	16
770	17	1.6	12.0	15	15	18
Bulls (growth + maintenance, moderate activity)						
660	20	2.0	12.3	27	23	34
880	24	2.3	15.4	23	23	43
1,100	27	2.4	16.5	22	22	48
1,320	26	2.2	16.1	22	22	48
1,540	28	2.4	17.0	23	23	50
1,760	23	2.0	12.8	19	19	41
1,980	25	2.2	13.9	21	21	44
2,200	27	2.3	15.2	22	22	48

Adapted from *Nutrient Requirements of Beef Cattle*, National Research Council, 1976 by M. A. Semlek, N.Y. S. College of Agriculture and Life Sciences.

1. Mature cow, 1,000 pounds, gestation. Note: average body weight should remain constant unless cows are thin, in which case they should gain.

A. 16 to 18 pounds mixed hay
B. 30 pounds corn silage, 4 pounds mixed hay
C. grazing crop residue (stalks, diverted acres, meadow aftermath), 2 pounds legume hay
D. 10 pounds straw or chaff, 8 pounds legume hay
E. pasture

2. Calving through rebreeding, 1,000-pound mature cow. Note: the cow should gain approximately 1 pound per day. The best quality hay or pasture available should be used at this time. Early spring pasture, however, may need supplementation with extra grain because of its high water content (cows won't be able to consume enough of it to meet their requirements). These feed levels should be reached by the thirtieth day postpartum.

A. good quality legume–grass or improved pasture (may need 3 to 5 pounds of grain in the early spring), free choice mineral mix
B. 10 pounds legume–grass hay, full feed corn silage, 40,000 I.U. Vitamin A, free choice mineral mix
C. 5 pounds corn or 7 pounds oats, full feed legume–grass hay, free choice mineral mix

3. End of breeding to weaning

A. good quality pasture, free choice mineral mix

If pasture is poor, creep feeding of calves starting at six to eight weeks is recommended, using primarily oats or oats and corn if corn is cheap (oats and corn can be fed whole). Calves will consume up to 500 pounds of grain up to weaning time.

4. Replacement heifers, weaning to breeding, gaining 1 to 1.5 pounds per day.

A. 25 pounds corn silage, 1 pound legume hay per 100 pounds body weight, 20,000 I.U. Vitamin A, free choice mineral mix
B. 10 pounds legume–grass hay or 20 to 25 pounds legume silage, 1 pound oats or ¾ pound corn per 100 pounds body weight, 20,000 I.U. Vitamin A, free choice mineral mix
C. good quality pasture, free choice mineral mix

5. Bred yearling heifers or bred heifers carrying second calf

A. grazing diverted acres or crop residues, 1 to 1.5 pounds of a 40 percent protein supplement, 20,000 I.U. Vitamin A, free choice mineral mix

B. 19 pounds legume–grass hay, free choice mineral mix

C. 30 pounds corn silage, 6 to 8 pounds legume–grass hay, 20,000 I.U. Vitamin A, free choice mineral mix

Mineral mixes with hay or pasture can be trace mineralized salt plus dicalcium phosphate. With rations low in calcium (corn silage), feed limestone should be supplied in addition.

Raising Feeder Calves

The first step in raising your own beef is getting the calf to raise. The calf may be a by-product of your dairy cow, it may be from your own beef herd, or it may be a purchased feeder calf. The next step for novices is to avoid naming it, unless you name it something like Steak or Ham Berger. It's much easier to sit down to a dinner of Steak than to a dinner of Spot or Frisky, Algernon, or Cutie Pie. This fiat only applies, of course, until you become hardened. Our current beefer goes by the name of Roger, and because ol' Rog has been such a pain in the neck (for a variety of reasons), rather than feeling sentimental and misty-eyed, I actually look forward to sitting down to a Roger barbecue. However, I do know people who

Fig. 6-4. Parts of a Steer

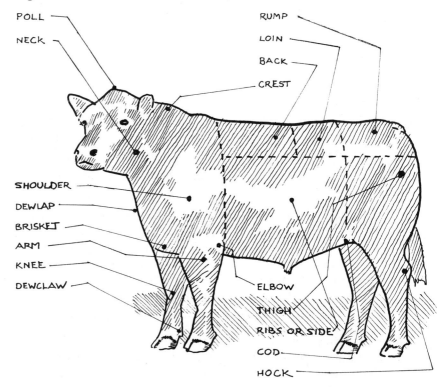

had gone to all the trouble of raising calves and finishing them, only to fink out at the *moment critique* because they just couldn't face eating little Inky. Avoid the problem from the start: don't get attached to your steer.

The breed of animal you get, if you're buying one, will largely depend on what's available in your neighborhood. Try not to get hung up on breed; a crossbred beef or dairy animal will be just as good for your purposes, in all likelihood, as a straightbred one. Unless you're experienced in looking at cattle (knowledgeably, that is, not just with a lot of fence-leaning in your background), try to buy a feeder from a local small-herd owner rather than through an auction. The following are points to be aware of when selecting a feeder:

1. Health: buy only a vigorous, healthy-looking calf. Danger signals include listlessness; rough, shaggy coat; gaunt appearance, sunken eyes, nasal discharge, and lack of appetite.

2. Weight: buy a calf weighing at least 400 pounds to reduce the possibilities of the diseases often associated with lighter weights. Yearlings should weigh at least 550 pounds. Weight can be estimated by using a weigh tape, available at many feed supply stores, or a tape measure, converting the measurements as shown in Table 6-3. Of course, the most accurate measure is a scale; by all means use it if the seller has one available.

3. Age and size: calves twelve to fifteen months old, weighing less than 500 pounds, have been nutritionally mismanaged and will probably remain small permanently. Age can be estimated by:

Table 6-3. Estimated Liveweight of Beef Cattle Using Heart Girth Measurements

HEART GIRTH MEASUREMENT*	ESTIMATED LIVEWEIGHT	HEART GIRTH MEASUREMENT	ESTIMATED LIVEWEIGHT
(in.)	(lb.)	(in.)	(lb.)
46	308	64	775
48	349	66	785
50	392	68	847
52	438	70	912
54	486	72	979
56	537	74	1,049
58	591	76	1,119
60	649	78	1,195
62	667	80	1,271

* Measured just behind forelegs

A. Length of tail: if the tail reaches the ground, the calf is probably over twelve months old.
B. Size of head: if the head is disproportionately large in com-
parison to the body, the calf is small for its age.
C. Teeth: a reliable, but difficult and impractical method of determining age.
D. Birth date: you can always ask when the calf was born. Who
knows? Maybe you'll get a straight answer.

4. *Price*: shop around and find out what calves are going for;
prices are quoted by the pound or in dollars per hundredweight.

5. *Sex*: steers or heifers are your best bet, although beef from
bulls under fifteen months old is perfectly good.

Bulls do grow faster than steers or heifers, but handling them
can be a problem (how's that for understatement?). If you're interested in raising a steer, but the seller has only bull calves, either:
(1) look elsewhere, (2) ask the farmer to castrate the calf, (3) get a
vet to castrate the calf. (Bull calves really should have been castrated well before weaning.) Heifers are a good choice because although they grow more slowly than steers, they are slaughtered at
lighter weights (they are finished at weights 150 to 200 pounds
lighter than steers) and often are cheaper as well (60 to 80 percent
the cost of a steer).

When and What to Buy

Following the usual beef cattle patterns, there are a few different ways one can buy a feeder. One could buy a weaned calf in
the fall; winter it on roughages and necessary supplements; graze
it the following spring and summer, with grain finishing the last
two months or so (if desired); finally slaughtering it in the fall,
when it's fifteen to eighteen months old. Or one could buy a yearling
in the spring, feed it and/or pasture it over the summer, and slaughter it in the fall. If you're lucky, you might find spring-weaned
feeders of about six months old; feed them over the next six months
into the fall, and then slaughter them as yearlings. The system you
choose depends on your resources. If you haven't got silage (and
what small producer has?) and you have no place to store the three-
quarters to one ton of hay the beefer will eat over the winter, your
best bet would be to keep the animal only during the pasture season
(assuming you do have pasture). If, on the other hand, you have
pasture, hay, and suitable shelter, wintering the animal is not such
a problem. Consider all your resources (access to hay, feed, pasture,
fencing, shelter, labor, etc.) and how much more you might be
willing to invest before deciding what type of beef raising is most
suitable for you.

Housing for feeders can be quite simple. It can, in fact, be nonexistent. If the animals are to be kept only over the mild summer months, a few trees to act as shade and as shelter from driving rains are all that are necessary. For animals kept through the winter, somewhat more shelter is better: a three-sided shed is fine; a well-sheltered woodlot is adequate if it's all you can manage. If the cattle are confined, allow at least fifty square feet per head.

Feeding

The feeding system you follow will depend on the resources you have available. In many countries, grass-finished cattle are the norm; grass finishing is used where grain is scarce or costly, forages are plentiful, and slaughter of an older animal is acceptable. Grain finishing results in faster gains and is lower in labor. Where grain is relatively cheap, therefore, and labor is expensive, the cost of gain might be lower when a grain ration is used instead of a forage-based ration.

If you intend to utilize forages to their maximum, be aware that the forage must be of good quality for it to put any weight on your beasts. Depending on the quality of the forage, some supplementation may be necessary.

For pasture feeding, the only equipment needed is a good fence (woven wire, barbed wire, or single strand electric), a watering area (a pond or tub), a salt block, and shade. For winter feeding, a hayrack, supplement feeder, source of fluid water, and mineral box or salt block will be needed.

Table 6-4 presents some guideline rations, which are based on The American Way of Beef—that is, relatively fast gains, acquisition of the proper finish with a certain amount of fat, and slaughter at a fairly young age (twelve to eighteen months old as compared to two or more years old in other countries). Animals fed according to these rations should gain 2 to 2.5 pounds per day. If pasture is being used, the hay may be omitted from the ration; depending on the quality of the pasture, a beef steer will need from one to three acres for the season.

When you first get your animal, feed it only good quality hay and delay grain feeding for a few days. Start feeding the grain gradually, increasing the calf's allowance by about a half a pound per day, until the full ration is reached after two or three weeks. (Just as a rule of thumb, cattle will consume about 2.5 percent of their body weight per day in dry feed.) If you switch rations at any time, do so gradually.

Remember that these are guideline rations which can and should be adjusted to the individual animal and situation.

Table 6-4. Guideline Rations

RATION No.	WEIGHT OF YOUR ANIMAL		
	400–600	600–800	800–1,000
1		lb./hd./day	
1 lb. ground ear or shelled corn per 100 lbs. body weight	4–6	6–8	8–10
Good quality hay or haylage (at least 50% alfalfa) fed free choice	4–8	8–10	10–12
Free choice trace mineralized salt supplement	.15	.2	.2
Total	8–14	14–18	18–22
2			
1.5 lbs. of shelled or ground ear corn per 100 lbs. body weight	6–9	9–12	12–15
Nonlegume hay such as timothy or brome fed free choice	2–5	5–6	6–7
Free choice trace mineralized salt supplement	.15	.2	.2
Total	8–14	14–18	18–22
3			
2 lbs. of shelled or ground ear corn per 100 lbs. body weight	8–12	12–16	16–20
1.5 lbs. of 40% protein supplement (or equivalent)	1.5 lbs.	1.5 lbs.	1.5 lbs.
Nonlegume hay such as timothy or brome (optional)	1–3	1–3	1–3
Free choice trace mineralized salt supplement	.15	.2	.2
Total	8–14	14–18	18–22
4			
1¾ lbs. of oats per 100 lbs. body weight	7–11	11–14	14–18
Good quality hay or haylage free choice	2–4	2–4	2–4
Free choice trace mineralized salt supplement	.15	.2	.2
Total	8–14	14–18	18–22

From M. A. Semlek, N.Y. S. College of Agriculture and Life Sciences, Cornell University, Ithaca, N.Y.

If the grains being fed do not contain a vitamin or mineral supplement, provide a free choice mineral mix. For rations containing mostly hay or pasture, a mix of 50 percent dicalcium phosphate and 50 percent trace mineralized salt is adequate. For rations low in calcium (i.e., corn silage), 40 percent feed limestone, 30 percent dicalcium phosphate, and 30 percent trace mineralized salt provide the necessary minerals. If the grains *do* include a mineral supplement, provide simply salt-free choice.

Always provide plenty of fresh water; your animal will need from five to fifteen gallons per day, depending on its size and the environmental conditions.

Optimum Slaughter Weights

Optimum slaughter weights are based on the breed and sex of the animal, with an eye to the proper finish. The question of finish is a rather hazy one for the small producer. It is said that animals slaughtered at lighter than their optimal weight will yield less juicy and flavorful meat, since they will not have acquired the desired degree of fatness. The amount of fat, however—although critical to the commercial producers, as it affects the grade of their cattle and consequently the price received—may not be as important to those raising beef for the family (especially if they plan to grass finish their beef in order to avoid the costly grain needed to put on the finishing fat). Also, fatness, particularly the intramuscular fat (marbling), is not necessarily correlated to tenderness.

Slaughter Weights

Heifers and some steers, British breeds (Angus, Hereford, Shorthorn)	800–950 pounds
British breed steers and crossbred British European breed steers and heifers (crosses on Charolais, Simmental, Limousin, etc.)	950–1,100 pounds
Large breeds (European breeds and large dairy crossbred animals such as Angus Holstein, Hereford Holstein, etc.)	1,100–1,250 pounds

Although weight is used as a major criterion in deciding when the animal is ready for slaughter, the amount of fat covering, as already mentioned, is also used. As the animal deposits fat, the rump, brisket, back, loin, flank, and cod, or udder, become fuller and thicker compared to other areas of the body, and as the animal reaches the desired finish, these areas will bulge somewhat.

Experienced cattle people can closely gauge an animal's weight just by looking at it; novices had better resort to more mundane methods. Weigh tapes for beef cattle—tape measures that correlate heart girth and weight—are available in many feed supply stores. In lieu of a weigh tape, a regular tape measure can be used, and heart girth (as measured just behind the front legs) can be converted to weight by using Table 6-3. Of course, a livestock scale is the most accurate weight measure, but unfortunately they are not usually readily accessible.

New Ideas for the Small Producer

Before getting into slaughtering methods, yields, etc., some attention should be given to the question of whether the entire system outlined up to now is really appropriate for the small stockholder and what alternatives there may be. The major drawback of the beef system as outlined is one of scale. What works well for keeping and raising 200 head of cattle is *not* 200 times one head; it is not usually feasible to simply scale down large enterprises to small ones, keeping the same techniques, standards, etc. Instead, a *new* system, geared to the small-scale stock raiser, should be devised. This is not to say that the techniques applicable to large-scale operations should be ignored, but rather, that they should be modified into new systems.

The biggest problem for the small-scale beef producer is the cost of the cow; maintaining that brood cow all year, just to produce one calf, is a mighty expensive proposition, and coming up with ways to reduce that cost is a worthy goal. (Actually, the cost of the cow is the biggest cost for the large-scale, commercial cow–calf raiser, as well.) Then, the type of animal produced and how it is produced should be geared to your specific needs. For example, if you are raising just a few feeders, one for yourself and a couple for the freezer trade (private individuals buying a half or whole carcass directly from the producer), your standards may be different from those of the industry. You may not be looking for maximum weights; weights of about 800 to 900 pounds—as opposed to 1,000 to 1,100 pounds—are more salable to private consumers who want only a family-sized, freezer load of beef. Freezer-trade consumers may also be happy with less of a finish on the animal; many producers raising their animals for this trade aim only for a low choice finish. Rapid, early gains also are not necessary—an animal eighteen to twenty-four months old may be just as acceptable as one twelve to fifteen months old. As you can see, the usual industry standards simply might not apply to all situations; following are some alternative methods for small-scale producers.

Hand Rearing Calves

The simplest way to cut down on the cost of maintaining a brood cow is to eliminate her. In many areas, dairy calves or dairy beef crossbred calves are available (many dairymen breed their first-calf heifers to beef breeds). Several of these calves can be bought and raised on milk replacer. If the calves are bought in the fall, they are fed milk replacer for two to three months and are weaned to grain and hay for the next few months; then they are turned out to pasture in the spring when they are about six months old—just when they're ready for pasture and the pasture is ready for them. At this point, the calves can be fed grain right through the pasture season to get them to approximately eight hundred pounds by late fall, at which time they are slaughtered. (Of course, they can, if necessary, be wintered over another season.) This system requires a good bit of labor and management to keep the calves healthy, especially in the early weeks. Time, however, is usually what the smallholder has and is willing to invest; by spending the extra time on the calves in those first few months, a considerable savings on the cost of producing beef calves can be realized.

Two Calves

Most cows, beef or dairy, produce more than enough milk for their one calf. Why not try to tap this resource? Many researchers are working on experiments involving the fostering of a second calf onto a newly freshened beef cow. When the cow calves, her calf is removed immediately, then two new calves (such as a Holstein bull calf and a beef calf) are returned to her, in the hopes that she will adopt them both. There are problems in getting the cow to accept the new calves, though, particularly if her own calf is returned along with the alien; again, time on the part of the manager can often overcome the difficulty. If you are there when the cow calves, remove her calf before she even gets to smell it. Have the alien calf in another pen. Rub the birth fluids and membranes over the alien calf, and put the two calves together. Bring the cow to the calves and wait to see what happens. If she doesn't accept both calves, try tying her and helping the calves to nurse. With effort on your part, both calves should be accepted in a few days.

Nurse Cows

This system avoids the problem of fostering an alien calf onto a newly calved cow. It also has a lot of possible variations which you can toss around in your mind in reference to your own situation.

In this case, a cull dairy cow or a crossbred dairy beef cow is used. A dairy cow is often a better choice, however, because her maternal instincts are better developed, she's easier to handle, she produces more milk, and she's usually easier to come by. So, you get a dairy cow—a smaller one is best, but a Holstein will do—and foster two to four calves onto her (depending on the cow's milking ability) at any point during the first couple of months of her lactation. Many cows will quite readily accept the calves, but some may need to be tied for the first several nursings. The basic idea now is that the cow raises these calves to weaning age and then is sold; several feeder calves, therefore, are raised on a cow kept only half a year.

If the cow and calves are bought in the fall, they are fed and housed over the winter; the cost of feeding the calves, while they're nursing, is really the cost of feeding the cow. In the spring, the calves are weaned and the cow is sold. The calves are then put out on pasture and fed for slaughter in the fall. On the other hand, the cow and calves could be bought in the spring. The cow would then nurse the calves while she is on pasture (being fed any supplementation necessary to keep her production adequate) and be sold after the calves are weaned in the fall. The calves would be raised through the winter on harvested feeds, and then be pastured in the spring and finished the following fall. If purchased in the fall, the cow must be maintained through the winter on harvested feeds while the calves nurse, but the calves can then make use of the spring pasture. If the cow is purchased in the spring, the cow's feed costs are reduced because she nurses the calves while *she's* on spring pasture, but then the calves must be kept through the winter on harvested feeds (unless they are sold as feeders in the fall). Clearly, some pencil pushing is required before you can decide which system would be the cheapest in your area and with your conditions.

A major requirement of this system is easy access to calves and easy buying and selling possibilities for cull dairy cows—which is just the situation in many dairy areas. Time and management are also demanded, but possibly less than for hand rearing. The economics of it can be quite attractive, in spite of the high labor costs; you can, theoretically, wean as many as four calves off one cow kept and fed for half a year. The cow herself only costs the difference (if any) between her buying and selling price, plus the cost of trucking and handling.

Breed

Don't automatically accept the beef breeds as the best animals to use; instead give some thought to the type of operation you're involved with and to the type of animal that will best suit it. A small, well-producing brood cow is more economical to keep than a big, heavy one. Many people claim that the ideal beef brood cow is a Jersey-Angus cross: a small, easy calver and a good milker. This crossbred cow (a Jangus, Jergus, Angsey?) can be bred back to an Angus or to a larger beef breed if a larger calf is desired. The problem with using Jergus cows for the large-scale producers is where do you get them? But for the person keeping only a few brood cows, this is not as big a problem; usually just a bit of asking around will turn up a few of these crossbred cows. Of course, a nice side advantage to a Jangus is that it's a good dual-purpose animal which might fit in just perfectly for some smallholders.

If you plan on purchasing a feeder calf, don't rule out the possibility of a straight Holstein or Brown Swiss, especially if they are easily available at the right time of year. As stated previously, dairy animals can produce excellent and economical beef.

All the previous ideas are aimed primarily at making the smallholder think about alternatives to the traditional, large-scale beef operations. Of course, in your particular situation, the traditional methods might be best. Just give it some thought.

Slaughtering

Approaches to slaughtering beef cattle are the same as for any meat animal. You can (1) do it all yourself at home; (2) have someone slaughter the animal on the farm, and then have it cut up elsewhere; (3) bring the live animal to a custom slaughterhouse. With each of these methods, be sure the animal has been handled gently before slaughter; if you have the heart, though, it is best to withhold all feed and water for twelve to sixteen hours beforehand.

The first approach requires facilities and knowhow. (See the references in Chapter 5 for information on home butchering.) It also requires cool temperatures or a cooling room so that the carcass can be hung at the necessary 30 to 45 degrees. Meat slaughtered and dressed on the farm can be used for home consumption only. If the animal is slaughtered on the farm but cut up at a custom plant, technically a "farm dressed certificate" must accompany the carcass. Slaughtering the animal on the farm eliminates the need for trucking the live animal, but you do need to have a space suitable for the butchering—and a beef steer is a *big* animal. Also, unlike pigs, cattle really are relatively easy to truck.

If the entire process is done at a slaughterhouse, butchering and cutting, the meat still cannot be sold; for the meat to be sold legally, it must be slaughtered under federal inspection and stamped accordingly.

Generally, if you're interested in selling beef, the best way to do so is on the hoof; that is, the buyer buys the live animal. Then, if you want to, you can make arrangements for the slaughter of the animal for the new owner and deliver the meat as well. Of course, if the meat has been inspected, any method of selling it is legal.

Once the carcass is ready, it should be aged in cold storage for a week to ten days to help tenderize it. Should the carcass have a heavy fat cover (.4 inches or more over the rib), it can be aged up to three weeks for additional tenderness and flavor. With less fat covering, however (less than .2 inches of outside fat), the aging process should be kept to three to five days.

Yields

What can you expect from your 1,000-pound steer? A lot depends on the actual carcass weight, the amount of fat, and the way it is cut up. Dressing percent in beef ranges from 58 to 65, which means you'll have a carcass of 580 to 650 pounds. Depending on the amount of fat trim, a carcass of that weight should yield about 400 to 500 pounds. (See Chapter 5 for details on the beef carcass.)

Diseases

Proper feeding, handling, and management of animals and prompt treatment of problems and injuries when they arise can greatly reduce the incidence of disease among your cattle.

1. *Shipping fever* is a type of pneumonia which results from severe stress. Poor appetite, fever, and rapid respiration rate are indicative of this disease. As it progresses, a thick nasal discharge and labored breathing develop. The disease may be fatal or can turn into chronic pneumonia. Vaccination is used as a preventative, but proper handling of animals (gentle handling; providing hay and water to new animals but starting grain slowly; avoiding stressful climatic conditions) is even better. Should you suspect that an animal has shipping fever, isolate it and contact the vet.

2. *IBR* (infectious bovine rhinotracheitis, or red nose) is a viral disease of the upper respiratory tract. Symptoms include high temperature, a clear discharge from nose and eyes, and redness, cracking, and peeling of the nose; abortion and secondary pneumonia may be complications. Vaccination of calves at six to eight months provides a three-year immunity to IBR. No treatment for the disease is presently available.

3. *BVD* (bovine virus diarrhea) is another viral disease, whose symptoms include an elevated temperature; ulcers of the tongue, lips, and mucosa of the mouth; and diarrhea, which may or may not be flecked with blood. The diarrhea is so severe that affected animals dehydrate rapidly and their eyes appear sunken. BVD may also cause abortion. Vaccination, given to nonpregnant cattle, will provide immunity.

4. Leptospirosis—see Chapter 3.

5. Blackleg occurs primarily in cattle under two years old and is characterized by raised temperatures, muscle swelling (particularly on the hind legs), bloat, and quick death; the disease must be treated early if death is to be prevented. Blackleg is caused by bacteria that can live in the soil for years, so regular vaccination is necessary in infected areas.

6. Salmonellosis is thought to be transmitted initially through fecal contamination of feed by rats and mice. Milk fever, poor appetite, bloody diarrhea, high fever, generalized infection, and death can be signs of this disease. There is no vaccination for it, but some antibiotic treatments are successful. Consult your vet.

7. Foot rot (necrotic pododermatitis) is caused by extremely hardy bacteria which enter the foot through cracks and abrasions. Lameness and swelling are symptomatic. Foot rot occurs most often during pasture months and should be treated by trimming away all dead tissue on the walls and sole of the hoof. If the underlying tissue appears healthy, penicillin will usually stop the source of the disease. Prevention of foot rot rests with management, as the bug thrives in damp conditions. Keeping cattle out of wet areas and maintaining proper drainage around heavily used areas, therefore, will reduce the incidence of this disease.

8. Bloat—see Chapter 3.

9. Indigestion—see Chapter 3.

10. Traumatic gastritis, or "hardware disease," is the exclusive property of cattle, who because they basically swallow without chewing, often ingest hardware—nails, wire, etc.—along with their food. (Now you know why your mother told you to chew your food.) These bits of stuff settle in the reticulum and can remain harmless, or they may prick the stomach wall, resulting in pain and lack of appetite, or they may actually puncture the wall, causing still more severe problems, including death (a pretty severe problem by any standard). Good management is the key to prevention: don't leave junk and wire around, and if you see a nail, pick it up.

11. External parasites include face flies, horn flies, grubs, and lice and are prevalent during pasture season (except for lice, which increase in the fall). Horn flies and face flies breed in fresh cattle manure and are generally controlled by the use of back rubbers, dust bags, sprays, or feed additives. (Back rubbers and dust bags are effective, of course, only if the animal uses them.) Ask at your feed store or cooperative extension office for recommended local practices.

12. Brucellosis—see Chapter 3.

13. Vibriosis—see Chapter 3.

For calf diseases, see Chapter 3.

References

Beef Production Reference Manual. Cornell University Extension Publication. Ithaca: New York State College of Agriculture and Life Sciences, 1977.

HOBSON, P. *Raising a Calf for Beef.* Charlotte, Vt.: Garden Way Publishing, 1976.

JUERGENSON, E. M. *Approved Practices in Beef Cattle Production.* Danville, Ill.: The Interstate Printers and Publishers, 1974.

PRESTON, T. R., and M. F. WILLIS. *Intensive Beef Production.* Oxford and
New York: Pergamon Press, 1970.

7
SWINE

Pigs are a riot. I know that's not a very scientific reason (although true) for raising pigs, but fortunately there are loads of other reasons for keeping pigs on a small holding.

In many countries pigs are kept as backyard scavengers who clean up garden surplus, miscellaneous roots and nuts, kitchen wastes, and dairy by-products, efficiently converting these inedibles (for humans) into highly edible pork. In the old days in this country, pigs were called "mortgage lifters" because they were such an economical source of meat. Today, many people in rural areas still keep a few pigs, raising them by a blend of old and new methods: making maximum use of food wastes and by-products, as was done in the old days, but also feeding some grain, which forms the total diet for the modern commercial pig. For the backyard pig producer, waste food is easy to come by once the domestic supply of garbage has been exhausted. Around here, all summer long on the grocery stores' "throw-out" day, there are always people ready to haul away crates of wilted lettuce, old squash, past-expiration-date yogurt, etc. to take back and toss to their eagerly waiting pigs. (If your eyes are bigger than your stomach at a restaurant, be sure to ask for a "piggie bag" for the leftovers.)

Then, too, the home pork producer ends up not only with economical pork but also with delicious pork. People who have never raised their own meat find it difficult to believe that there is any difference between commercially produced meat and homegrown, although they'll readily extol the virtues of garden fresh vegetables over store-bought. But there is a difference—a very noticeable one—and once you've consumed your homemade pork, you'll be very hesitant to buy the commercial variety. Besides, what better use is there for all those baseball bat-sized zucchinis (which you can't even pay someone to take out of your sight) than to turn them into pork?

Easily and economically fed, user of wastes and garbage, simply housed and cared for: these are all sound reasons for keeping a pig. But then there's the fun, which is something a large-scale swine producer may never notice, but which the couple-of-pigs owner can delight in. I think of my pigs running through the tall grass of their pen in the cool of the morning, or wallowing ecstatically in their mud hole during the heat of the afternoon. I think of how they play with my dog, follow me around (if I let them), and play games with each other. They're clever, intelligent, personable little critters, fun to have around, and if you're not careful, easy to get attached to. (I must add a note here for my friend Bob, who is probably saying, "Yeah, and the fun of chasing them through the field in the pitch dark when they break out in the middle of the night. Or the fun of rounding them up when the neighbor calls to tell you your pigs are in his corn." I can still hear Bob during that nocturnal pig roundup, moaning, "Five Acres and Slavery—I've had it with getting back to the land!" It's just a fencing problem, Bob, read on.)

Large-Scale Enterprises

Even though keeping backyard pigs is quite different from running a swine enterprise, a lot of the management principles are the same, just on a smaller scale. Obviously, a two-pig operation is going to be more labor-intensive than a large-scale one—you'll be using water buckets instead of automatic waterers, and so forth—but the basic principles used in handling one sow and her litter are pretty much the same as for fifty sows. The same is true of finishing three pigs instead of three hundred: the major difference is one of scale. The pig is still a pig, who has certain basic requirements, which can be met in a variety of ways. It's a good idea, therefore, for the potential pig owner to have at least some notion of the form the swine industry takes in this country.

There are two separate types of swine production: purebred and commercial. Commercial is further divided into feeder pig, growing-finishing, and farrow-to-finish production. Although purebred operations are not important in terms of the number of head produced (less than 1 percent of the eighty million pigs produced annually are purebreds), they are extremely important in supplying the foundation breeding stock used in commercial enterprises. It is the purebred breeders who lay the groundwork for change and genetic improvement in swine. Because detailed, accurate records must be kept on purebreds, more labor and better management skills are required of the operators.

Commercial hog production supplies most of the country's pork and generally relies on crossbred animals (often the offspring of a purebred boar mated to a sow of another breed or a mixed breed sow). The feeder pig type of commercial operation involves the care of a breeding herd, raising piglets to approximately six to ten weeks or forty to fifty pounds, and selling them as feeder pigs (pigs to be fattened and finished). It requires fairly expensive equipment, and is considered labor- and management-intensive.

Growing–finishing operations, on the other hand, are the simplest type of swine enterprise, requiring the least amount of equipment and managerial skill. In this type of enterprise, feeder pigs are purchased, fed to market weight, and then sold. No breeding or farrowing problems need be dealt with; all you need is a place for the pigs and plenty of feed. Although some grower–finishers feed their pigs on pasture, there has been a significant move toward confinement finishing, where pigs are kept in lots on concrete or slatted floors. This kind of feeding is attractive to commercial swine producers because of the faster rates of gain, the ability to handle more pigs, and the freeing of land for more valuable crops. The disadvantages include social problems among the pigs (fighting, tail biting), higher labor requirements, more parasitism, and problems with manure disposal. If land is available, however, finishing pigs on pasture is still considered economical.

Farrow-to-finish programs are just what they sound like: the whole shebang. This is still the most common type of hog production, and it obviously involves all the advantages and disadvantages of the other types of operations combined. The biggest advantage is that by this method the swine producers supply their own feeder pigs and can, therefore, control the quality of their market hogs by their choice of breeding stock. This system also provides the greatest profit potential per unit of production, although it requires a large capital outlay.

Breeds, Types of Pigs, and Selection

There are over eighty breeds of swine in the world, and more than twenty breeds are found in the United States. Table 7-1 summarizes some of the traits of the more popular breeds.

Pigs can generally be classified into one of two types: bacon or ham. Bacon pigs are longer and leaner (such as the Landrace), whereas ham pigs are stockier and squarer (such as the Duroc). The types of pigs to be found dominating the market at any one partic-

Table 7-1. Characteristics of Some Popular Swine Breeds

BREED	COLOR MARKINGS	EAR SET
Yorkshire	solid white	erect
Hampshire	black, with white belt around body at shoulders, including both forelegs	erect
Chester White	solid white	drooping
Berkshire	black, with white feet, face, and tail	erect
Duroc	red	drooping
Landrace	solid white	drooping
Poland China	black, with white feet, face, and tail	drooping
Spotted	black and white spotted	drooping

ular time depend on consumer demand. Pigs can be bred to respond amazingly quickly to market preferences. Compared with other farm animals, significant changes in body type can be effected rapidly. Gestation length in pigs is short (just under four months), gilts (young sows) can be bred at a relatively early age, and litters are large. Coupled with the fact that body type is highly heritable and is easily and cheaply judged, all this means is that when the market demands lard, the pig producers can quickly start turning out fatter, lard-type pigs. And, of course, when the market wants lean meat

Hampshire sow.
Courtesy: Hampshire Swine Registry.

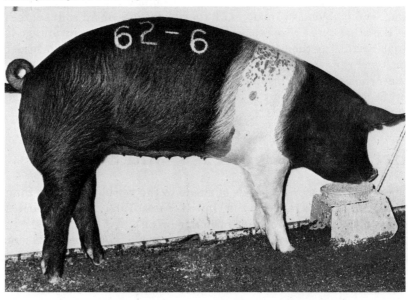

(as it does now), the producers can respond by selecting and breeding for a leaner type of carcass. Many commercial swine producers are now aiming for a lean, long pig with good ham, but also with good bacon (they want the best of everything).

When choosing a few feeder pigs to grow and fatten out over the season, the main thing the smallholder will look for is thriftiness and general body type. If, however, you plan on raising a gilt to breed, or are choosing a sow for breeding, you should be more concerned with the records and background information available on the pig. Buy breeding stock only from reputable dealers; auctions can be adequate suppliers of feeders, but you'd want more information on breeding animals than you can get from an auction. Ask the seller for records on the sow, or if buying a gilt, for records on her parents. Look particularly at litter size and number of pigs weaned. A sow that farrows ten and weans five isn't as good as one that farrows eight and weans eight (as long as you're sure management isn't the entire difference); don't buy a pig that weaned fewer than five pigs per litter. Weaning weight of the piglets is another factor to consider, as it reflects the nursing ability of the sow. Look at the animal herself; she should be moderately long, with squarely set legs, have a wide appearance from the rear, with plump hams and well-muscled hind legs set well apart. She should be healthy, vigorous looking, and free from genetic defects.

As mentioned earlier, feeder pigs can be bought from auctions, but it is slightly risky. It is less likely that you'll get a bargain price when buying from a private party, but by checking out where

Hampshire boar.
Courtesy: Hampshire Swine Registry.

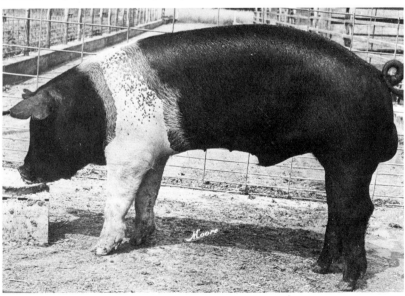

Chester White sow.
Courtesy: Chester White Swine Record Association.

the pig comes from and talking to the owner, you can be more assured of the health and the quality of the pig. I once bought a pig at an auction at such a bargain price I was sure it would die or not grow or something hideous. To everyone's surprise and pleasure, though, she did quite well; it is possible to luck out, but by and large, "you pays your money and you takes your chance" with auctions. It's not difficult to find people selling piglets in the spring;

Spotted barrow.
Courtesy: National Spotted Swine Record.

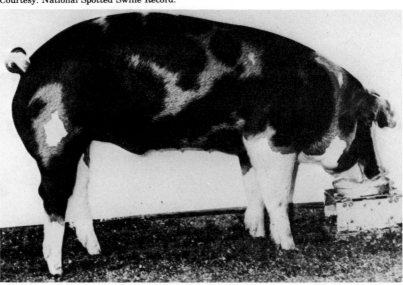

Landrace boar.
Courtesy: American Landrace Association, Inc.

newspaper ads, feed stores, or neighbors will usually put you in touch with a pig seller. Ask around a bit, though, and check the auction and market prices to establish the going rate for feeders. The pig market bounces around so much that it is impossible to give any estimate here of what feeders cost. Last year, for example, you could get an eight-week-old pig for fifteen dollars; right now, eight-week-old pigs are bringing as much as thirty-five dollars. Who knows what the market will be like a year from now?

Managing a Brood Sow

Shelter and Equipment

One of the many nice things about keeping pigs is that they are not very demanding when it comes to housing. A pen in the barn during the winter, and a pen with an exercise area, or a portable pig house on pasture during the summer is all that is needed. When owning just a sow or two, you can afford to be luxurious with their quarters, but the minimum size for a sow pen or movable house is seven by eight feet. The floor can be anything, but dirt is best. (Commercial piggeries, concerned with strict sanitation and manure disposal, use concrete or slatted wood, but for the one-sow owner, this flooring is not necessary.) For the portable pig house, just about anything will do, from a simple A-frame to a Quonset hut. Right now we're using the discarded wooden cap from a pickup truck, and it works great. Orient the opening of the shelter away from the weather, and check the drainage to be sure the inside of the house stays dry. Provide bedding, and as with any animal, keep it clean. Pigs are naturally very clean creatures, and given enough space, will choose one area of the pen for elimination, making pen cleaning a relatively simple business.

Of course, when keeping a sow, farrowing quarters in addition to the sow quarters will be required. Farrowing pens are discussed later in this chapter.

Although less important for nonpregnant gilts and sows, an exercise area is a must for pregnant animals. This can range from an eight-by-ten-foot outside area accessible from the pen, to a much larger fenced-in space, to actual pasture. If the animal is on pasture, be sure that shade is provided. Pigs cannot easily dissipate body heat by sweating, so if possible, in addition to shade, provide a wallow; the pigs will love it and you'll love it. They're so obviously happy and contented in their wallow that you can't help feeling good just watching them. For the opposite problem, cold, provide a lot of dry bedding—the sow will dig a nest into it. If the pig must stay outside during the winter, again, provide plenty of clean bedding and keep the shed dry and draft-free. Heaping horse manure against the walls of the shed will help enormously to retain the heat, and hanging a burlap bag across the doorway will help keep out the weather while allowing pigs to move in and out.

One warning about pens and huts in general, though: they must be structurally sound. Pigs are incredibly strong and will either crash through or systematically destroy anything even slightly flimsy. They're also incredibly itchy; they love to scratch against things, and 250 pounds scratching against a shed wall is a lot of stress.

The same is true of fencing: it must be geared against scratching on and rooting under. The usual method of discouraging pigs from rooting under woven wire is to run a single strand of barbed wire just underground below the fence. If your fence is good and tight, though, and sunk into the ground a couple of inches, escapes through rooting should not be a problem. (It also depends somewhat on the personality of the pig; some pigs are worse rooters than others.) By far the easiest fencing to use for pigs is electric. Two-strand electric fence works well (sometimes even one strand will do; we've always managed with just one), but unlike other livestock, pigs seem to know when the current is off. I don't know how they know, because they're not constantly touching it to check, but there's no doubt that they *can* tell. A pig farmer who uses electric fence said he thought the pigs could *smell* the current. They do have very sensitive snouts, and maybe if they can smell a truffle buried in the ground they can smell electricity in a wire. Who knows? In any case, don't take a full-grown sow and throw her out into a field fenced with electric wire unless (1) she came from a place that used electric fence, (2) you teach her about the fence in a smaller space first, or (3) you like chasing after sows.

However, once the pig knows you (and knows that you're the source of her dinner), even if she does get out, it probably won't be hard to get her back in again. (For this reason, keep your new pig confined for a few days so it can get to know you.) Also, if they have no reason to break out, your pigs probably will stay put regardless of the fencing. For our pigs we have quite a large, well-vegetated space fenced in with a single strand of electric wire. There have been times when I've gone up to feed, the fence current would be off, and I'd find a pig *chewing* on the wire. Obviously she knew the fence was off, but she didn't really have any reason to go anywhere.

Feeders and waterers, too, can range from the improvised (a dishpan) to the slick (commercial metal self-feeders). Whether you have trouble coming up with a feeder and waterer for your pig will again depend a lot on her personality. Most pigs will root under, stand in, and scratch on any feeder you might try, with some pigs doing more damage than others. But if the feeder is sturdy and either untippable (very large or heavy) or well anchored (with stakes driven into the ground), damage will be lessened. Damage in this case refers less to actual physical damage of the feeder and more to the waste of feed. Feed is too costly and valuable to be able to regard its waste casually. Commercial self-feeders, designed to hold a quantity of feed wastelessly (they have lids which the pigs lift up in order to eat), are expensive but can be worthwhile if you have a few sows. (These metal, lidded self-feeders are also good for rodent control.) I have seen homemade wooden copies of these commercial feeders, and they work well, although there is some risk of a bored sow chewing the feeder as well as the feed. When using an improvised feeder or trough, make sure the pigs have access to all sides, especially with sows on limited feeding; what is spilled will generally be cleaned up.

A large rubber bucket is perfectly adequate as a waterer. It must be tied securely in place, though, since in hot weather the pig

Fig. 7-1. Two Styles of Homemade Pig Feeders: A Hopper and a Trough

will delight in knocking over the bucket and flopping onto the wet spot. The waterer, whatever form it takes, must be capable of holding at least three gallons of water per pig; pigs drink a lot, and the water must be available at all times.

Feeding

Pigs are simple-stomached animals, and as such, must be fed carefully balanced rations containing good-quality protein. If you're keeping one or two sows, you will probably be buying commercial complete feeds. (Some sample rations, if you're making up your own, are shown in Table 7-2.) However, since feed represents about 75 percent of the total cost of keeping a pig, and since a sow can consume as much as two thousand pounds of feed per year (depending on the number of pigs farrowed and weaned and what wastes and by-products she may be getting), you want to be sure that you are not feeding any more feed, or any more-expensive feed, than is necessary.

As with other animals, the sow should be fed according to her stage of production, which over the year will be maintenance, gestation, or lactation, with a short period of prebreeding. If a sow is farrowing two litters per year, there will be, in fact, only a few weeks per year that she is not in gestation or lactation.

Maintenance If your sow is farrowing only one litter per year, there will be times when she should be fed only a maintenance ration. Four and one-half pounds per day of a balanced 12 to 14 percent

Table 7-2. Eight Grain Mixes for Sows for Hand Feeding during Gestation or for Self-Feeding during Lactation (in pounds)

Mix	1	2	3	4	5	6	7	8
Ground yellow corn	49.5	40.0	73.2	75.5	74.5	67.0	78.4	73.0
Ground oats	25.0	15.0	—	—	—	10.0	—	—
Alfalfa meal	10.0	15.0	10.0	10.0	10.0	7.5	5.0	10.0
Wheat middlings	—	15.0	—	—	—	—	—	—
Tankage or meat scrap	7.0	6.0	6.0	10.0	—	—	—	—
Soybean meal	7.0	8.0	10.0	—	10.0	10.0	10.0	15.0
Steamed bone meal	1.0	0.5	—	0.5	—	—	—	—
Salt (plus trace minerals)	0.5	0.5	3.0	0.5	0.5	0.5	0.5	0.5
Dicalcium phosphate	—	—	0.3	1.0	0.4	0.4	0.4	1.0
Meat and bone meal	—	—	—	—	3.8	3.8	5.0	—
Ground limestone	—	—	—	—	0.3	0.3	0.3	—
Vitamin premix	—	—	—	—	0.5	0.5	0.5	0.5

ration will be adequate for sows around 250 pounds. Alfalfa hay is well utilized by sows, as is pasture, and can form up to 15 percent of the ration. If good hay or pasture is provided, or plenty of scraps or dairy by-products are available, the amount of grain fed daily can be reduced somewhat; be aware, however, of the protein quantity and quality of the diet.

Prebreeding Assess the condition of your sow a month or so before breeding season. She should be thrifty and healthy looking, neither fat nor thin. The number of ova shed during the heat period is greatly influenced by the condition and nutritional level of the sow; females that have been improperly fed or underfed will not produce a normal number, resulting in a small litter. Flushing the sow has been shown to be effective: a week to ten days before breeding, start increasing the sow's feed level so that she will be in a gaining condition at breeding time. Keep feeding at the higher level until she is bred, then switch to a gestation diet. The flushing increases the number of ova shed, thereby potentially increasing litter size. A sow that is too fat at breeding time, however, may have difficulty conceiving.

Gestation The biggest problem in feeding the sow during gestation is preventing her from becoming too fat. Sows that are overweight at farrowing tend to have small, weak pigs, so one must be careful to limit the daily intake of the sow to just what is required and no more. Exercise is very important at this time and will help to keep the sow in condition. The ideal situation is to have the pig on pasture, or at the very least, access to a large, grassy yard. A good pasture will supply much protein as well as vitamins and trace minerals; a large dirt area with some vegetation will help insure an adequate supply of trace minerals and vitamins, although it will not supply much protein. If pasture is impossible, good quality legume hay should be supplied to the sow. Good pasture for your sow can be quite economical because it reduces the amount of grain required; three pounds of grain per head per day for sows on pasture is equivalent to four pounds per head per day for confined sows.

You *do* want the sow to gain some weight during gestation, however, both to support the development of the fetuses as well as to put the sow in good condition for lactation, and possibly, early rebreeding. As a general guide, sows should gain sixty to seventy pounds during gestation, and gilts should gain eighty to ninety pounds. Since most of the fetal growth occurs during the last third of gestation, the larger part of the weight gain should occur during that period. (Note: it has been found that high levels of energy intake early in gestation contribute to increased losses from early embryonic death, so overfeeding should definitely be avoided at that time.)

Depending on the age (gilt or sow), size, and condition of the pig, three to five pounds per day of a complete concentrate feed containing 15 to 16 percent protein is satisfactory. The amount fed should be less during the first two-thirds and increased during the last third of gestation. Protein quality is especially important during the last third, as is good forage and exercise. In cold weather, increase slightly the quantities fed to allow for the increased requirement of the sow for maintaining body temperature. A ration utilizing dairy by-products would be corn fed free choice, legume hay or good pasture, plus five pounds buttermilk or skim milk.

Three to four days before farrowing, add 20 to 25 percent wheat bran to the ration to add bulk and reduce the chances of constipation.

Lactation Feed requirements for lactation are much higher than for gestation: three to four times higher, in fact, with a high TDN requirement and a high protein level. Because gestation and lactation rations are so different, the changeover period is critical and the change should be made gradually. Withholding all feed, but keeping a fresh supply of water available, for twelve to twenty-four hours after farrowing will help prevent udder trouble at the start of lactation. Milk production gradually increases after farrowing, peaking at four to five weeks. Daily feed intake is increased from approximately four pounds per day at the onset of lactation to full feeding (ad lib, or approximately twelve pounds per day) by the tenth day postpartum. (A rule of thumb is to feed a quantity that is 3 percent of the body weight of the sow, or four to five pounds of feed per day for the sow plus an additional one pound of feed per nursing piglet.)

For the sow to produce to her capacity, she must have a high-energy, 15 to 16 percent protein ration. You may be able to buy a suitable feed ready-mixed, or you may have to have one mixed up especially for you. Some sample mixes for lactating sows not on pasture are shown in Table 7-2. These rations are suitable for feeding the sow during gestation as well, as long as you are limiting the sow's intake (hand feeding) rather than letting her eat as much as she wants from a feed hopper (self-feeding). If the sow is on *good* pasture, the pasture can substitute for the alfalfa in the ration.

Breeding a Sow or Gilt

Successful breeding begins with successful heat detection. Since the owner of a sow or two will not, in all likelihood, own a boar (who makes heat detection *much* easier), some indication is needed of when to bring the sow to the boar. (Some boar owners will bring the boar to you, if you have adequate facilities, and leave him for three weeks, thereby insuring a breeding and simplifying the sow owner's job of heat detection.)

The length of the estrous cycle in gilts and sows (from the start of one estrus to the start of the next) is approximately twenty-one days, with a normal range of eighteen to twenty-four days. The signs of early estrus are a response to the boar (if there's one around), mounting of other females and allowing other females to mount her, but *not* allowing the boar to mount. The vulva will look swollen, and a vaginal discharge will appear as estrus progresses; distinct restlessness and frequent urination accompany these signs around the second day. A commonly used check is the application of pressure to the sow's back. If she's in estrus, she will assume a rigid stance, referred to as lordosis. Signs of estrus will persist for three to four days, but the sow will mate only for two to three days. Occasionally, these signs can be intensified or brought out by allowing the sow to sniff a rag that has been wiped over a boar; if she is in heat, the sow should react to the boar's odor.

Ovulation occurs on the second day of estrus, with the ova shed over seven or more hours. The number of ova shed during estrus increases with each successive estrus the sow goes through, up to the fourth or fifth cycle. Gilts bred at their first estrus, therefore, will have a smaller litter size than if they are bred at a later estrus. Also, most first-litter gilts farrow fewer pigs the first time than they will subsequently. (Gilts raised for breeding should be kept at about 250 pounds and can be bred at eight or nine months. Feed requirements for gilts in gestation are slightly higher than for sows to allow for the gilt's own growth as well as the fetal growth.)

Especially when breeding a gilt, but even with mature sows, the relative weights of the boar and the female should be kept in mind. Not surprisingly, a sow can have her back broken during mating with a too large boar. If there is too much disparity in their sizes, the use of a breeding crate is recommended.

Gestation length is approximately 114 days (I always remember three months, three weeks, three days), so the owner can choose between trying for either one or two litters per year. With one litter a year, farrowing can be timed to coincide with the pasture season and mild weather, reducing feed and supplemental heat requirements. Where winters are cold, two farrowings per year will inevitably result in one occurring during the cold weather, necessitating careful management. In the Northeast, many people aim to have their pigs farrow between February and April to have feeder pigs ready for the spring market. Some will then rebreed their sows between April and July, with a fall farrowing around August to November. Sows are then rebred October to January, resulting in February to May farrowings. However, fall pigs sold as feeders from October to January command a somewhat lower price than the spring pigs. Whether to have one or two farrowings, and at what time of year, should be based on your assessment of the market and economic situation and the availability of labor and feed.

Farrowing

Conscientious care of the sow and piglets at farrowing is critical; providing the proper quarters for farrowing and baby pigs, and being there when the sow farrows, will greatly reduce problems and losses. A commercial swine producer once told me that he felt that being present at farrowing saved at least one pig per litter. That may not sound like a lot, but if you've got one hundred sows, that's one hundred pigs. Even with just one sow, one piglet represents 10 percent of a ten-pig litter and can be a significant part, if not all, of your entire profit.

Farrowing quarters can range from a pen with a farrowing crate, as used in commercial enterprises, to a large, well-bedded pen equipped with guardrails, to a hut on pasture. Farrowing crates—large boxes (twenty-four inches wide by seven inches long)

A sow nursing her litter.

Fig. 7-2. Farrowing Pen Equipped with Guard Rails. A heat lamp
in the far corner makes a hover for the piglets, providing
both supplemental heat and extra protection against accidental crushing.

that hold the sow during farrowing—are used to prevent sows from
rolling on the piglets and crushing them (called overlaying, or
pancake disease). But with more pen space available to the sow
than is normally provided with the crate system, crushing is less
likely to occur. Even in a large pen (larger than eight by ten feet;
pens should be long and narrow rather than square), guardrails
should be installed. These are rails eight inches above the bedding
and eight inches from the wall, that provide an escape space for the
little piggies. Then there is the farrow-in-the-field option, which is
really just a warm-weather alternative. A shelter with a lot of
bedding must be provided, and the sow will then build a nest,
farrow, and wean the little ones herself. The advantage to this
method is the utterly minimal labor input; the disadvantage is the
probable weaning of smaller litters because of death losses. However, if labor is at a premium and little pigs aren't, you may choose
this method (which is really no method at all, since it is basically
allowing nature to take its course). For the person who wants to be
more sure of his pig crop, however, a pen with guardrails is the best
bet; most small-scale operators are not so cramped for space that
they need to use crates, or so tight for labor that they can't put the
time into farrowing.

In addition to ample space, bedding, and guardrails, the farrowing pen must be equipped with a brooder, or hover, to provide
supplemental heat for the piglets. Baby pigs cannot regulate their
body temperature well and so are extremely sensitive to chilling.

Soaking up the sun; but beware of sunburn on young white pigs.

A temperature of eighty-five to ninety-five degrees should be maintained in the pigs' sleeping area for the first three days after farrowing. The heat can be decreased gradually as the piglets get older, depending on ambient temperatures; watching the behavior of the pigs will help you decide when and how much to decrease the heat. For heat you may simply suspend a heat lamp in a protected corner of the pen at a height that will provide eighty-five degrees at the floor. To prevent burns, suspend the lamp about twenty-four inches above the floor if possible.

A pig brooder, or hover, is a more enclosed box, with a ten- or twelve-inch opening for the piglets; with this device, a one hundred-watt lightbulb may provide enough heat. A hover space serves a function in addition to the provision of heat: it draws the pigs away from the sow while they are not nursing into a safety zone. Since at least one-third of all baby pig losses is attributed to crushing, you want to do all you can to prevent it. If the sow is farrowing during the winter, extra heat will benefit her, as well; the ideal temperature range for farrowing quarters is fifty-five to sixty-five degrees.

The sow should be brought into her farrowing pen three or four days before she's due; a first-time gilt, who might be more nervous, can be brought in sooner. Once she's in the farrowing pen, wash her off and reduce her feed intake, switching to a laxative ration as outlined earlier. If you know the date the sow was bred, use that to estimate her due date. Otherwise, watch for the signs of the big moment approaching. Twenty-four to forty-eight hours before farrowing, the udder will fill with milk, and an actual milk flow will start about twenty-four hours before parturition. As far-rowing time approaches, the vulva will dilate and the tailhead will sink. Six to twelve hours before, the sow will appear nervous, may chew things, will urinate frequently, and will keep pushing around her bedding. A half hour before the first pig arrives, the membranes should rupture, causing a discharge from the vulva.

Piglets may be born head first or feet first, and usually the sow has no trouble at all getting them out. However, because piglets are often born with the membranes still surrounding their heads, it is a good practice to be present at farrowing to clear their faces to prevent suffocation. As each pig slithers out, pick it up, wipe the mucous from its nose and mouth, and give it a shake to encourage its first breath. Tie and cut the umbilical cord, leaving one to two inches; then dip the cord in an iodine solution. As with other live-stock, the navel cord in the newborn pig acts as a prime entrance site for bacteria, and failure to disinfect the cord at birth can result in (among other things) a disease known as navel ill, which causes lameness or death.

At this point, many large-scale operators also carry out other management practices: they'll record the sex and birthweight of each piglet, notch the ears for purposes of identification, and clip the needle teeth. Needle teeth are four pairs of very sharp teeth, two on each jaw, which if not clipped can cause damage to the sow's udder or to the other pigs as they all play among themselves. It's a simple procedure which even the small-scale pig raiser may con-sider. A clipper can be bought from most supply houses or farm stores for a few dollars. The piglet's head is held rather like a pistol, with the holder's thumb behind the head and the index finger inside the mouth across the back of the jaw. With the teeth exposed in this way, you can clip the tips of the teeth off, being careful not to loosen the base of the tooth or leave jagged edges. It may sound difficult, but it's really quite simple once you get the holding posi-tion, and it takes only seconds per pig.

Another procedure involves the prevention of anemia in baby pigs. Iron does not cross the placenta easily; the newborn pig, there-fore, is born deficient in iron, or anemic. To top it off, sow's milk is quite low in iron, so the deficiency cannot be corrected with nursing, and piglets must be given iron in some other way. For sows on pasture, this is not an issue; the soil contains enough iron to prevent the deficiency. If the sow and litter are in a pen, a large box of dirt can be provided for the pigs to root in and chew through; there's no guarantee that the pigs will be getting enough iron, but it is better than nothing and is certainly simple. Other techniques include the intramuscular injection of an iron dextran compound (following the manufacturer's directions) directly into the little pigs; swabbing the sow's udder daily with a saturated solution of ferrous sulfate (one pound of ferrous sulfate in one gallon of water), so the piglets get some each time they nurse; or the administration of any of a variety of iron pills or pastes to each pig individually.

Once you have completed your care of each newborn piglet, and have seen to it that heat is available and the piggies are comfortable, pay some attention to the sow; be sure she has completely delivered and has fully cleansed. If she seems to be straining, but nothing is happening, consult your vet.

Baby pigs are very active, curious little beasts and will quickly find their way to the milk bar. Occasionally a dumb pig will start sucking on something other than a teat; just show it to a teat—the reward of sucking a teat as opposed to an ear will be enough to teach it the proper place to nurse. Once a piglet has established a teat position, that's it. Within twenty-four to forty-eight hours, each pig has its own personal milk source, untouchable by other piglets. This can cause a bit of a problem if there are more pigs than nipples (sows have six to twenty nipples, with an average of twelve to fourteen) or if a pig gets stuck with a dry or low-producing teat. Little piggies in the nest do not agree, and they won't share. If there are more than enough teats to go around, try to switch the shortchanged pig to a better nipple as soon as possible. If there simply are not enough nipples (or too many pigs), you'll have to hand raise the extra pigs or foster them onto another sow. (Fostering is difficult, at best, and impossible if you've only got one sow.)

Raising Orphan Pigs

To transfer orphan pigs to another sow, the foster sow must have farrowed recently, within a few days of the transfer in fact, because unused teats dry up fast. Separate the sow from her litter for a bit; then mix in the new pig, sprinkling disinfectant or some other odor-disguising material over all the pigs. Then return the sow to the litter and hope for the best.

Barring the availability of a recently farrowed sow, you'll have to raise the orphan by hand. As long as you have the time, the task is not too difficult.

House the baby pig in a wooden box or draft-free space and provide heat, maintaining a temperature of eighty-five to ninety degrees during the first week. If the pig is kept warm, it can survive about twenty-four hours without milk; but the first feeding really should be given as soon as possible, and certainly within the first twelve hours. If the pig can get colostrum for its first feedings, great. If not, your job of pulling the pig through is going to be a little harder.

Feed the pig as often as possible. When piglets suckle naturally, they nurse about once an hour around the clock. The feeding
schedule you set up can approximate this one as much as you like,
but you must be prepared to feed a minimum of four evenly spaced
feedings per day. You can bottle-feed, using a regular baby bottle,
or teach the pig to drink from a bowl. Use a shallow bowl and
gently push the pig's snout into it; baby pigs generally can learn to
drink this way by the second feeding or so.

To concoct a good sow milk replacer, thoroughly mix one egg
yolk with one quart of cow's milk. Except for iron, this mixture will
give the piglet a balanced diet. One-eighth teaspoon of ferrous
sulfate should be added to it to make up for the iron deficiency (or
you can use injectable iron as stated previously). The milk–egg mix
can be used as the total food source for the first three to four weeks
of the pig's life.

The major problem encountered in raising pigs by hand is
scouring, which is usually caused by overfeeding or feeding at irregular temperatures. Warm the milk and feed several small feedings per day to reduce the chance of scours. Keep everything clean.
Should loose feces occur, reduce the amount of milk by at least half
for the next couple of feedings. Start by feeding the pig very small
amounts (a couple of ounces), gradually increasing the allowance
as the pig seems to be able to handle it.

There is another problem in raising an orphan pig: it can
easily become a pet, especially if, for the convenience of those frequent feedings, you've kept it in the house. Baby pigs are *really*
cute, and once they've made it through the first couple of weeks,
quite easy to take care of. I once raised a baby pig in the house—
one luckily destined for pethood at a summer camp. She quickly
learned to drink out of a bowl and was paper-trained faster and
younger than any puppy I've ever had. She chose the spot, we put
paper there, and she never went anywhere else. Because she was
so well trained, she had free run of our cabin until she joined, at
about five or six weeks, the rest of the gang in the outside pens. She
was a friendly, responsive, and altogether delightful little pet. However, if your pig is destined for the table, beware of giving in to its
cuteness. Also, although having a little pig follow you around is
cute, having a 250-pound sow follow you around can be a nuisance.
While still young, the piglet must be taught to live in the pigs'
quarters.

Losses of Baby Pigs

In addition to stillbirths, which may be as much as 3 to 8
percent of a litter, many piglets die because of poor management or
low birth weights; by far the greatest rate of mortality is during
the first twenty-four hours. Death by crushing can be avoided by

Fig. 7-3. A creep allows feeding of the piglets without hogging by the sow. The feeder in Fig. 7-1 is converted into a creep by the addition of a lower board, which blocks off the sow.

providing a large farrowing pen with guardrails, or if necessary, a farrowing crate. Chilling is easily avoided by the use of heat lamps or other supplemental heat sources. Being present at farrowing will prevent the suffocation of piglets in the placental membranes. The chance of infectious disease, such as navel ill and scours, is reduced by disinfecting the navel cord at birth and maintaining a clean, sanitary environment. If you lose piglets for any of these reasons, you know you have to sharpen up your management practices.

Lighter pigs have a much poorer chance of survival than their heavier siblings, so piglets under two pounds at birth should be watched closely—and if they do not do well or if they are *really* small, raise them by hand. The weight of piggies at birth is inversely related to the litter size: the bigger the litter, the smaller the pigs in it. Within the litter, first-born piglets tend to be heavier than the others; being bigger, they also tend to get the best nipple position (front teats produce more milk than rear ones), reinforcing the difference. Beware of great disparities within the litter that might need correction.

Remember that the prenatal nutrition of the sow or gilt affects the size and viability of the piglets: a good, healthy litter begins with good care of the dam.

Raising Piglets

Piglets will start nibbling at dry food by the time they're ten days to two weeks old, and at this point a creep ration is usually

Fig. 7-4. An Outdoor Creep Area. The small opening into the pen allows only the piglets access to the feed.

supplied. A creep feeder is an enclosure built into the corner of the pen with an opening of such a size that the little pigs can run in and out freely to get to the feed, but the sow can't get to it at all. If the sow and litter can be put on pasture once the pigs are two to three weeks old, a creep can be built outdoors, again making an enclosure or portable pen (pallets are good for this purpose) that only the piglets can get into. When designing a creep, allow at least one foot of feeder space per four pigs, and set the edge of the feeder no more than four inches off the ground.

Feed used for creep feeding must be of high quality. Pelleted rations are best, but ground grain can be used as long as it is not ground to the point of dustiness. Until they reach twenty-five pounds, the pigs should get a 20 to 22 percent protein feed; from twenty-five to fifty pounds—after which they will certainly be weaned—an 18 percent protein feed is sufficient. Some pig starter (creep) rations are shown in Table 7-3. For piglets on *good* pasture, corn (coarsely ground or shelled) or a mixture of corn and oats, plus a protein supplement (soybean meal, meat scrap, fish meal, or dried dairy products), both fed free choice, is suitable.

Keep the feed fresh (put out the amount eaten in one day, rather than dumping a lot and leaving it for several days) and provide plenty of fresh, clean water at all times.

Another preweaning practice is the castration of male pigs. One simply cannot produce meat for human consumption from an intact male: boars produce a scent which imparts an irradicable (and awful) smell and taste to the meat. Castration, therefore, must

Table 7-3. Grain Mixes for Creep Feeding (in pounds)

Mix	1	2	3	4	5
Ground yellow corn	68.6	58.0	50.2	54.0	80.0
Soybean meal (50%)	18.5	15.0	22.5	23.6	12.0
Dried skim milk	—	15.0	—	—	—
Meat and bone scrap	2.5	—	—	—	3.0
Condensed fish solubles	—	2.5	2.5	—	—
Fish meal	1.0	—	—	—	1.0
Dried whey	2.5	—	15.0	15.0	—
Stabilized fat	—	2.5	2.5	—	—
Cane or beet sugar	—	5.0	5.0	5.0	—
Cane molasses	2.5	—	—	—	—
Alfalfa meal	2.5	—	—	—	2-4
Calcium carbonate (38% Ca)	0.5	0.5	.5	.5	—
Dicalcium phosphate	0.9	1.0	1.2	1.2	—
Iodized salt	3.5	3.5	3.5	3.5	—
Vitamin premix	0.1	0.1	0.1	0.1	—
Trace mineral mix	0.05	0.05	0.05	0.05	—

be done, and it is best done by the time the piglets are three weeks old; at this age there is less shock to the pig, and consequently there is almost no setback period during which the pig fails to gain. The procedure is not a difficult one, and provided proper sanitation is maintained, infection is uncommon, but it definitely should not be attempted by a novice until he or she has watched an experienced person perform the operation. A clean scalpel, razor blade, or sharp knife is the only equipment needed, in addition to a disinfectant. While one person holds the pig up by its hind legs, gripping firmly and ignoring his screaming (since nothing has been done yet, he is in no pain), the other makes an incision through the scrotum. Either one incision between the testes or two incisions, one directly over each testicle, can be made. Be careful not to cut the testicle itself but only through the tissues surrounding it. Squeeze the testicle out through the opening, and scrape or crush and cut the cord (scrape rather than slice the cord to minimize bleeding). Spray the area well with disinfectant and return the pig to a *clean* pen: the male pig is now a barrow, not a boar. Observe the pig over the next few days to be sure the wound is draining well and is not becoming infected. You should realize that although the pig is screaming like a banshee while being held, as soon as you put him down he'll quit his yelling. Pigs *hate* being off their feet; and based on the noise they make, anyone would think you're torturing them cruelly every time you pick them up. However, from what I've seen, the little pigs are more concerned with getting back on their feet than with what's going on between their legs, so you just have to block out that awful shrieking and not let it make you feel terrible.

Weaning

Weaning is the next big step in the pig's life, and it can occur at the time you decide (three to eight weeks) or at the time the sow decides (eight to twelve weeks). It is common practice to wean pigs at five to eight weeks, but if you are raising only one litter per year, and especially if your sow and pigs are on pasture, you can let the sow wean the pigs herself. The only real disadvantage is that because the sow is nursing for a longer period of time, she will require more feed. Also, it is not feasible to let the sow wean the pigs if you plan on rebreeding her for two litters per year. (A sow cannot be rebred until the piglets are weaned.)

By six weeks baby pigs should be eating solid food well, so weaning at this time is not difficult. To cut down on the sow's milk flow, reduce her concentrates a few days before weaning. Then, if possible, move the sow (rather than the piglets) to new quarters. Should her udder become unduly congested, she can be returned to the piglets for a few minutes every second day until she dries up. (Now you can see the value of teaching your sow to be moved easily.)

At this point, the pigs should weigh about thirty pounds and should continue to get a high quality, high-protein (18 percent) starter ration, fed ad lib. Provide a minimum of one foot of feeder space per three piglets, and at least seven square feet of pen area per pig.

Early weaning—at three to five weeks—is practiced by some commercial swine operators. The advantage is primarily the savings in feed for the sow. However, the likelihood of heavier, more uniform pigs at market time (six to eight weeks) also makes early weaning attractive. Another advantage often claimed is that it is possible to rebreed the sow earlier, but in fact most sows cannot be successfully rebred until thirty days postpartum; this advantage, therefore, is dubious. The disadvantages to early weaning are many: decidedly superior management is required; the higher quality ration needed is expensive and may offset any savings on the sow's feed; and the pigs do not get to nurse during the peak of the sow's lactation. If pigs are to be weaned early, however, they should weigh at least twenty-one pounds at five weeks, fifteen pounds at four weeks, or twelve pounds at three weeks. The younger they are, the warmer their quarters must be: sixty degrees at five weeks; seventy degrees at three weeks. Provide at least six square feet of floor space per five-week-old pig, and four square feet per three-week-old pig, with a foot of feeder space per three to four pigs, maximum. Pigs weaned early must get a very high-quality 20 percent protein ration—often called a prestarter—up to the time they weigh twenty-five pounds.

Raising a Weaner: Growing–Finishing Pigs

Getting Your Pig(s)

Whether you've raised a pig to weaning size (thirty to forty pounds) off your own sow or you've bought a weaner pig, the process of getting it to slaughter weight is the same: feed it. If you've bought a pig, though, there might be a few preliminary items you'd want to be aware of.

It's upsetting for a pig to be moved, and the more it's been trucked around the more stressed it will be. A pig bought from an auction may have been severely stressed indeed, as it has been exposed to many other pigs and possible diseases. (A pig bought out of a neighbor's litter, however, will not be in this situation.)

To get your newly purchased pig off to a good start, be sure to have a quiet, dry, warm, draft-free, and well-bedded pen, complete with feed and water, all set up and ready for his arrival. If you think the pig was really stressed, you might want to provide medicated drinking water and feed with antibiotics for the first few days (ask at your feed store for medicated feed). Otherwise, just see that the pig is treated calmly. Establish your feeding routine right

One management technique is to let nature take its course. In left photo, a tropical farmer leaves his sow and her litter out to scavenge with the other animals; the sow weans the litter herself. Right photo, pigs on pasture.

away and stick to it; move slowly, and talk to the pig—it will be to your advantage in handling if the pig is not afraid of you.

If you're raising more than one pig and they're to be housed together, be sure they're both about the same size. Actually, it's a good idea to raise two pigs together; they'll eat more and grow faster. Two pigs are not much more work than one, and two can be quite economical, especially if you've got a lot of garden surplus. (The idea that someone else might get it makes that rotten old squash you throw in the pen much more appealing.) Homegrown pork is an easily marketed item, often commanding a premium price, so raising two pigs and selling one can significantly offset the cost of raising your own. Keeping two pigs, however, does require that more attention be paid to the strength of your facilities: two scratching pigs exert twice as much stress on buildings, fences, etc. as one.

The usual pattern is to buy a weaner in the spring and butcher it in the fall. This method makes maximum use of garden surplus and avoids the difficulties of caring for an animal during the winter. You can generally figure on keeping your pig for four to six months (a lot depends on how it's fed) so aiming for an October or November slaughter means getting the pig in May or June.

Housing and Equipment

One of the beauties of keeping an animal for the warm months only is the simplicity of housing: all that's needed is basic protection against sun and wind. A three-sided shed in a well-fenced field is great, and a pen in a barn with access to a yard is also good. As with sows, providing a wallow is not a necessity, but it *is* a nice amenity which will make those hot summer days much more bearable for the piggies. For you straight economics-minded folks out there who may scorn providing amenities for pigs' sakes, remember that a comfortable, happy pig is an eating pig, and to keep the pig eating and growing is the name of the game.

Minimum space requirements for confined pigs are: fifteen square feet per pig of feeding space in addition to sleeping space of five to six square feet per pig up to 75 pounds, seven to eight square feet per pig from 75 to 125 pounds, and nine to ten square feet per pig to market weight.

Fencing was discussed previously, but a few extra points can be stuck in here. As mentioned earlier, woven wire fencing, if tight and in good condition, is excellent for pigs but is very expensive. If you decide to use electric fence instead, teach your pigs about it before depending on it totally. Keep them penned in their shed or in the barn or wherever for the first several days, letting them get used to you and teaching them that yours is the hand that feeds them. Then electrically fence in a small space, or use a few wire strands in addition to a board across an open doorway, or use some setup that will allow a backup method of capture should the pigs escape. (The best backup is having the pigs know that this is home; if you're certain they won't go anywhere even if they *do* get out, you can put them right into their yard.) Make sure that they all touch the wire at least once, and make sure the charge stays on at all times. Once they've lived with the training wire for a day or so, you can move them into their yard or field. Be sure that the wire is visible in the field in the same way as in the smaller space (we tie white rag strips on the wire at close intervals).

There is nothing more pathetic than someone going out and eagerly buying a pig, bringing it home, and then losing it because of faulty fencing and lack of orientation time. Just last year our neighbors bought three pigs and put them right out in a yard fenced with old woven wire and boards. Well, it took about three hours for the pigs to find a hole in the fence and make their escape. They hung around our pigs for a while but were so shy of people that they couldn't even be tempted with the all-time pig treat: milk. Eventually they just disappeared from the area altogether. Who knows what happened to them? Maybe some hunters that fall got a pig surprise while stalking their deer. In any event, it was seventy-five dollars right out the window (or out the pen) for our neighbors. If the pigs had been kept more tightly confined until they had felt at home, they probably would have been easy to catch if and when they had escaped.

Although in all likelihood the small-scale stockkeeper will not be dealing with minimum space requirements, be aware that the minimum standards are twenty-five square feet of pen space per pig.

In addition to shelter and a yard or pasture, the only other items required are a feeder and waterer. Fresh water must be provided daily in a sturdy, untippable trough or bucket capable of holding at least three gallons of water per pig. If using a trough, allow one and a half feet of space per pig; if using buckets, provide one bucket per one or two pigs.

Any indestructible container able to withstand the rigors of being rooted under and stood in, and which can hold the necessary quantity of feed, is suitable for a feeder. Many people use a fifty-five-gallon drum (or any other size drum) cut in half lengthwise (crimp the edges, of course) and anchored into the ground with stakes. Friends of ours use a heavy soapstone sink. We've used metal troughs, unanchored; the pigs pushed them around but never spilled and wasted the feed. In fact, making sure the pigs can get around to all parts of the feeder will help insure that any feed spilled will be eaten up. Pigs vary in their rootiness, but I have found that if they have plenty of other things to keep them busy (books, radio, real roots), they'll not bother much with destroying the feeder. Confined pigs get more bored and so present more of a challenge to the designer of a homemade feeder. If you're going to feed food dry, a heavy wooden feeder works well, but it is unsuitable for wet feed. Allow plenty of feeder space for your pigs to insure that each one is getting its fair share; one pig will tend to be dominant, so watch to be sure that he can't hog all the food (so to speak).

Feeding

Feeding your pigs to make the fastest, yet the most economical, gain is the big trick to growing-finishing pigs. Also, because what you feed affects the carcass composition, feeding becomes more than just dumping in a bucket of potato peels.

The first thing to realize is that growing pigs, being simple-stomached, high-producing animals (growing and fattening), need a high-energy, high-quality protein ration. As the pig gets older, however, the amount of protein required decreases since more and more of the feed is being converted to fat. Table 7-4 shows the protein levels required in the diet at the different stages of the pig's growth. Should your pigs be on pasture—decent pasture, not a weed patch—the protein content of the feed can be reduced by 2 percent. From ten to twenty pigs can be kept per acre of good pasture, depending on whether the pigs are being full fed or limited fed. (Some sample rations are shown in Table 7-5.) Prices of different feeds, of course, will ultimately dictate ingredients in the rations. (Most small producers will do best to buy a commercially mixed feed if a suitable one can be found.) Pigs fed diets low in protein from weaning to market weight produce carcasses with less lean and more fat; so if you're formulating your own rations and have a tendency to skimp on the more expensive ingredients, be aware of the possible effects.

The next big question is how much to feed, which is affected by two major types of feeding systems: full feeding, which is feeding the pig as much as it will eat, and limited feeding, in which the manager feeds a certain amount each day. Full feeding is advantageous to the person feeding grain only: the labor is less and the pig grows faster. However, if you are feeding a fair amount of table scraps, by-products, pasture, or roughage, limited feeding is necessary to encourage the pig to eat foods other than grain. The pigs may grow somewhat more slowly by this system, but if labor costs are not high, the gains will be more economical and more efficient. (Efficiency of gain refers to the number of pounds of feed required per pound of gain.)

Certainly most people raising a couple of pigs would do well feeding leftovers, weeds, etc., in addition to grain. Old cucumbers, pea vines, broccoli plants, corn cobs and husks, wilted lettuce—anything—just dump it in the yard and the pigs will love it (although I have to admit that this year even my pigs got tired of

Table 7-4. Protein Levels Required in Feeds

Stage of growth	creep	starter	weaning to 75 lbs.	75-125 lbs.	125 lbs. to market (220)	breeding herd
Percent crude protein	20	18	16	14	12	15

Table 7-5. Grain Mixes for Growing-Finishing Pigs (in pounds)

WEIGHT OF PIG (LBS.)	50-100				100-150				150-200			
	1	2	3	4	1	2	3	4	1	2	3	4
Ground yellow corn	73	74	59	78	79	81	65	83	85	87	65	88
Ground oats	—	—	19	—	—	—	19	—	—	—	19	—
Soybean meal	19	14	10	18	14	11	5	13	8	5	5	9
Meat scrap	—	5	5	—	—	3	5	—	—	3	5	—
Dehydrated alfalfa meal	5	5	5	—	5	5	5	—	5	5	5	—
Steamed bone meal	2	—	2	—	1	—	0.5	—	1	—	0.5	—
Dicalcium phosphate	—	2	—	3	—	.5	—	3	—	0.5	—	3
Trace mineralized salt	0.5	0.5	0.5	0.5	0.5	0.5	0.5	0.5	0.5	0.5	0.5	0.5
Ground limestone	1	—	—	1	0.5	—	—	1	0.5	—	—	0.5

From W. G. Pond, E. A. Pierce, Cornell Extension Bulletin 1045.

zucchini). If you have any surplus or sour milk, or any dairy by-products (whey, buttermilk, etc.) the pigs will be ecstatic—and fat; milk really seems to put the weight on. Leftovers from supermarkets and restaurants (your piggie bags) are also good. In addition to the scraps, you want the pigs to eat the vegetation in the yard or field, which they will do with amazing rapidity. On top of all this waste feed, give the pigs grain. (Table 7-6 shows pig performance and how much grain a full-fed pig will eat.) If you want to encourage the pigs to eat all these other things, though, feed only 75 percent as much grain as they would eat if full fed. If there's still feed in the feeder the next day, reduce the next feeding a bit until the pigs are cleaning up what is fed to them. (Unwittingly, my mother taught me the philosophy behind this method, only she applied it to people: when serving dinner, if there's a little left over, you know you made enough; if there's a lot left over, you either made too much or the food was awful.) It will take approximately 600 pounds of grain to get the pig to market weight.

Unless you are feeding *a lot* of dairy products or meat scrap, feed at the protein level recommended; in general, waste foods are an energy, not a protein, source. A good ration using dairy products is corn, barley, or wheat fed free choice, legume hay, and six pounds of buttermilk or skim milk. If good pasture is available, maintain the grain feeding, omit the hay, and reduce the milk to three to four pounds per day.

Grain can be fed either wet or dry. Many pigs seem to prefer their feed wet, but it's sloppier; the choice is yours.

Slaughter

The optimum slaughter weight for a pig is 200 to 220 pounds, which should be reached by the time the pig is six months old. Growth rate peaks at this weight, then declines, and further gains, which are increasingly fat gains, are less economical and efficient.

Table 7-6. Pig Performance, Full Feeding (in pounds)

LIVEWEIGHT	AVERAGE DAILY GAIN	DAILY DRY FEED INTAKE	FEED PER POUND OF GAIN
25	0.8	2.0	2.5
50	1.2	3.2	2.7
100	1.6	5.3	3.3
150	1.8	6.8	3.8
200	1.8	7.5	4.2
250	1.8	8.3	4.6

From W. G. Pond, E. A. Pierce, Cornell Extension Bulletin 1045.

You *will* end up with a heavier pig if you keep on feeding it, but it basically will be just a fatter pig. A 220-pound liveweight pig will produce about a 158-pound carcass, yielding approximately 110 pounds of retail cuts. Chapter 5 shows details of the pig's carcass.

How do you know when your pigs are approaching slaughter time? You could be arbitrary and go by age; a properly fed pig should reach slaughter weight by six months, but this figure is hardly accurate and could end up being quite uneconomical if your pig happens to reach slaughter weight younger. You could also arbitrarily use feed as a guideline; since it usually requires about 600 pounds of grain to finish a pig, you could just call it quits once the 600th pound has been consumed—also not a very accurate guide. Or you could keep putting off the decision until you can no longer justify saying, "Just one more bag of feed." The best way to judge, of course, is by looks and weight. A ready pig will *look* ready, complete with nicely rounded hams, a wide body, and a look of firmness about the flank, belly, and jowl; it will weigh between 200 and 220 pounds. In the absence of a scale, a hog weigh tape can be used easily to estimate the weight of the pig. (A hog weigh tape is available from The Highsmith Co., Inc., Fort Atkinson, Wisc.).

Once you've (reluctantly) decided slaughter day is upon you, how do you get the job done? Basically, there are three ways to handle it: (1) do it all yourself, (2) bring the pig to a slaughterhouse, or (3) have someone come to your place and slaughter the pig. I've never slaughtered a pig myself, and considering the cost of having someone else do it, I probably never will. For six to eight dollars you can have a cleaned carcass, which you can cut up yourself or have cut up by a butcher for about ten cents per pound. Bringing the pig to the slaughterhouse is fine with the proper trailer or truck but is a real hassle without. (Load pigs quietly and gently with a minimum of upset—guide them from behind with a stock cane, but *do not hit them*.) It is a lot easier to have someone come to your place and slaughter the pigs, if possible, but you must provide a place for the carcass to hang, and you must clean up afterwards; I think it's a fair tradeoff, though, for not having to move the pigs.

As mentioned, you can cut the carcass up yourself or have the butcher do it, and the same is true of smoking or curing. Morton Salt publishes an excellent booklet on home curing and cutting, which also includes explicit directions for home butchering. Many other publications covering home slaughter are available; a few are listed in the reference section of Chapter 5. One word on smoking and curing: many people automatically have the hams cured, but if you've never tried it, have at least one fresh ham. Fresh (uncured) ham is delicious, and it's a product rarely found in the supermarket.

Once you've got this lovely product—homegrown pork—all packed away and ready for use, do yourself a favor and don't over-cook it. People are so paranoid about trichinosis that they tend to cook pork to the flavor and consistency of cardboard. One nice thing about growing your own pork is that you know what the pig has eaten. Trichina are transmitted through pork, so if your pig hasn't eaten any, the chances of it having an infestation are almost non-existent. (The incidence of trichina infestation among grain-fed pigs is less than 0.12 percent.) Also, since most of you will freeze your pork, you should be aware that freezing pork for twenty days at 5°F (−15°C) or below destroys trichinae, as does cooking pork to an internal temperature of 171° (well before the cardboard stage).

Diseases and Parasites

As always, good sanitation and management are the keys to disease prevention; nutritional diseases resulting from poor feeding, parasite problems from poor management, and infectious diseases from poor sanitation are all avoidable. The careful owner of a pig or two is unlikely to see any of the diseases outlined, but it always pays to be aware of potential problems. Be ready to consult your vet; very often, prompt diagnosis and treatment can make all the difference.

Nutritional Diseases

Baby pig anemia is a fairly common problem among suckling pigs. Caused by an iron deficiency, it results in a lowered resistance to disease and a slower growth rate. In severe cases, pigs become unthrifty looking, with puffiness around the throat and possibly "thumps," a labored breathing pattern. Prevention of baby pig anemia is simple (and is covered in the discussion of farrowing).

Trace mineral deficiencies do not usually occur in pigs fed well-balanced rations, as most feeds contain adequate amounts of the trace minerals. But for those toying with mixing rations yourself or getting by with just oats or something, here's a list of what can result from lack of trace minerals:

iodine: hairless pigs at birth
manganese: slow growth and abnormally short legs and body
copper: anemia, rough hair coat, weak legs
zinc: parakeratosis (see below)
cobalt: slow growth and anemia

Parakeratosis is caused by a lack of zinc. The need for zinc is increased by high levels of calcium in the diet and by large proportions of soybean oil meal. (Soybean meal contains phytic acid, which aggravates a zinc deficiency.) Symptoms are a rough, scaly skin on the underline, eventually covering the entire body and retarding the pig's growth. The scaliness is easy to confuse with mange.

Parakeratosis is easily reversed by correcting the zinc:calcium ratio. A ration containing more than 1 percent calcium can cause the disease; 50 to 100 ppm (four ounces per ton of feed) of zinc carbonate or zinc sulfate will cure or prevent it.

Rickets can be caused by a calcium, phosphorous, or Vitamin D deficiency or by an incorrect calcium:phosphorous ratio. Because early correction of these deficiencies will prevent any permanent and serious damage, mild rickets (seen as a bowing of the front legs) should never be allowed to turn into severe rickets (permanently damaged legs and slow growth). To prevent rickets, the calcium:phosphorous ratio should be maintained at 1.5:1, and adequate Vitamin D should be supplied. Pigs allowed access to a yard should get ample Vitamin D from the sun, so that rickets should not be a problem.

Parasites

Internal parasites are really gross, and on top of being utterly disgusting, they can do significant damage, causing pigs to be unthrifty, stunted, and inefficient. I still say, though, that the best way to convince a livestock owner of the need for worming is to show a picture of the inside of a heavily infested animal—just be grateful it's been omitted here.

Roundworms (*Ascaris suis*) are the most common internal parasite, but there are many other types of worms that can be detrimental to your pigs. Treating pigs for roundworms involves the use of piperazine, sodium fluoride, dichlorvos, or levamisole, and is quite simple. Occasional toxicity may be encountered with the use of sodium fluoride, but any of these compounds used in accordance with the manufacturer's directions should be safe. Piperazine has an advantage in being able to be used either in dry feed (as the others are administered) or in the drinking water.

Sows should be wormed before being moved to their farrowing quarters to reduce the chance of infestation of the baby pigs. Pigs should be wormed when they are about six to eight weeks old and then again four to six weeks later. (If the pigs are in a clean environment, though, and are not crowded, the second worming can probably be dispensed with.) When buying feeder pigs, try to find out what worming they've had and treat them accordingly. When buying a pig from an auction, play it safe and worm it.

External parasites aren't a treat, either. The two big ones for swine are lice and mange-causing mites. Lice, which are bloodsuckers, are usually found on the flanks, shoulders, and backs of ears; eggs may also be seen on the hairs, close to the pig's body. The standard treatment uses organic phosphates, such as malathion, ronnel, and Co-Ral. Ask at your local feed store, but best of all, avoid the need for them by good sanitation. Mange mites burrow into the animal's skin and cause itchy, scaly patches. They can be treated like lice—and prevented like lice.

Infectious Diseases

In all cases, call your veterinarian.

Hog cholera, responsible for large annual losses in the industry, is caused by a virus found in the blood, body tissues, feces, and urine of infected animals. The incubation period is three to seven days, with the first symptoms—loss of appetite, fever, coughing, eye inflammation, and a reddish or purplish discoloration on the ears and underline—appearing around the fourth day. The disease is spread by direct contact, by exposure of susceptible animals to infected ones, or by the virus being tracked in on shoes, boots, etc. Unfortunately, there is no treatment, and the best prevention is sanitation. A national hog cholera eradication program is in effect, and some states have been declared hog cholera-free.

TGE (transmissible gastroenteritis) is a viral disease which primarily affects suckling pigs, although it can occur in pigs of any age. The incubation period is eighteen to twenty-four hours, and symptoms include scouring, vomiting, and dehydration. When occurring in pigs less than two weeks old, mortality may be as much as 100 percent. There is no effective treatment for TGE, but high levels of antibiotics in the feed reduce the danger of secondary complications. Again, sanitation, especially at farrowing, is the best preventative.

MMA (mastitis-metritis-agalactia complex) is a set of problems associated with reproduction. Mastitis is an udder inflammation, metritis is a uterine inflammation, and agalactia is a lack of (or poor) milk flow. The causes are unknown, but *E. coli,* staphylococcus, and streptococcus bacteria seem to be involved, as may a hormone imbalance.

Mastitis usually occurs soon after farrowing; the mammary glands become swollen, hot, and painful, and the piglets can't nurse. Fever, loss of appetite, constipation, and reduced milk flow are associated with mastitis.

Metritis also occurs right after farrowing, with symptoms appearing within one to three days. Elevated temperature, lethargy, loss of appetite, and a white or yellowish discharge from the vulva are the general symptoms, accompanied by a hot, swollen udder and a reduced milk flow. Baby pigs could easily become weak and die of starvation if the condition is not recognized and treated promptly.

Agalactia is most often seen in gilts and young sows, and the lack of milk or poor flow is frequently the only symptom. This disease is indicative of a hormone imbalance, requiring hormone treatment.

Good management will greatly reduce the chances of any of these conditions.

Erysipelas is caused by a bacteria found in the feces and urine of infected animals and is spread by ingestion of contaminated feed or water or by the entrance of the organism through cuts or scratches. The organism is hardy and can survive in the soil for years, so pastures, auction yards, etc. can be sources of infection for animals coming into contact with those premises. The symptoms are similar to those of hog cholera: loss of appetite, listlessness, and fevers of 106 to 108 degrees or higher. Red, diamond-shaped patches appear on the skin, and the chronic form is characterized by lameness and swollen joints. Mortality ranges from zero to 100 percent, depending on the severity of the disease.

Brucellosis is another bacterial infection, caused by the organism *Brucella suis*. Infected feed, water, or body discharges spread the disease. Abortion, stillborn or weak pigs, small litters, or sterility are the major symptoms, although some animals may be carriers of the disease without showing symptoms. There is no known cure. The best prevention is to buy breeding animals known to be free of brucellosis.

Leptospirosis is another reproductive disease resulting in abortion or weak or stillborn pigs, caused by the bacterium *Leptospira pomona*. Leptospirosis is the single most important cause of abortion in swine, and the disease is transmissible to humans. It is spread in the same manner as brucellosis, with the principal symptom being abortion, but it can be prevented through vaccination and antibiotics in the feed.

Mycoplasma pneumonia, caused by a mycoplasma, used to be called VPP, or viral pig pneumonia, when thought to be of viral origin. Under either name, the symptoms are coughing and a slow growth rate. Since it is a chronic disease, it often goes unrecognized. Some estimates indicate that as many as 40 to 60 percent of the swine in the United States have VPP. The only sure way to diagnose the disease is by examination of the lungs at slaughter, looking for the purplish, fibrous lesions typical of the infected animal. There is no known cure.

Atropic rhinitis (AR) also results in slowed growth rate and greatly reduced feed efficiency. The cause of AR is still unknown: some people argue for nutritional causes (related to a calcium deficiency), others for infectious agents (*Bordetella bronchiseptica*), and still others for environmental causes (high ammonia levels in the barn). The symptoms are sneezing and a bloody nasal discharge in young pigs. As the disease advances, the nasal turbinates in the snout become distorted, and the nose may end up crooked or folded-looking. It is thought that 10 to 20 percent of the swine in the United States have AR.

Note: Many of the diseases mentioned are spread through direct contact of an infected animal with a susceptible one, which

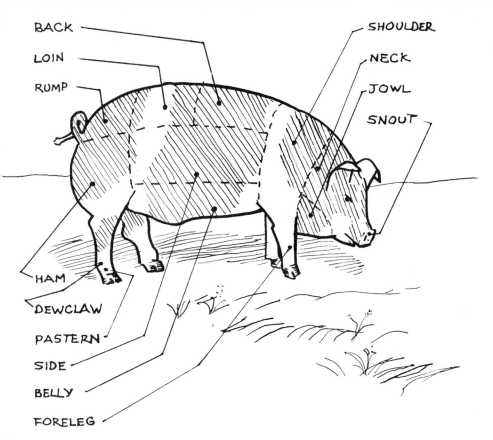

BACK

LOIN

RUMP

SHOULDER

NECK

JOWL

SNOUT

HAM

DEWCLAW

PASTERN

SIDE

BELLY

FORELEG

Fig. 7-5. Parts of a Pig

is why you are taking such a chance when you buy pigs at auctions. In the auction pens, young pigs come into contact with many other pigs, and not always under the best of circumstances. Remember, too, that people don't usually put their best pigs through auction but rather the smaller or weaker ones or those with some indication of problems. (I'm referring here to piglets being sold at general stock auctions, not to feeder pigs being sold at feeder pig sales.) It's easy to see why it's safer to buy a pig out of a herd where you can see the conditions and talk to the manager.

References

BELANGER, J. *Raising the Homestead Hog.* Emmaus, Pa.: Rodale Press, Inc. 1977.

POND, W. G., and J. H. MANER. *Swine Production in Temperate and Tropical Environments.* San Francisco: W. H. Freeman and Co., 1974.

POND, W. G., and E. A. PIERCE. *Successful Swine Production.* Cornell University Extension Bulletin 1045. Ithaca: New York State College of Agriculture and Life Sciences, 1974.

8
SHEEP

Sheep: (a) fluffy white things you count to put yourself to sleep? (b) cute little animals to use in nursery rhyme illustrations? (c) that which completes the pastoral scene on the English countryside? or maybe (d) all of the above—but not, not the livestock you expect to see on an American farm. Why?

Part of the reason is lamb consumption: although Americans consume over one hundred pounds of beef per person per year and more than sixty pounds of pork, the average American eats less than four pounds of lamb. If lamb makes you think of dried-up chops which absolutely *have* to be smothered in mint jelly to make them at all edible, then you probably think there's a good reason for this low rate of consumption, namely—lamb is awful. People in other parts of the world, however, would not agree. The English and the French, for example, are avid consumers of lamb and mutton, with marvelous recipes for various cuts. People of the Mediterranean countries, too, make terrific use of lamb: stuffed whole lamb, moussaka, and souvlaki, to name a few highlights of the lamb-eating world. So why do we fall so short in our appreciation of this truly delectable meat?

A major reason is that we cook it badly. Even those leathery little chops of childhood had potential at one time; they just should have been cooked about ten hours less than they were. Lamb should still be pink when done—not a uniform earth brown—especially leg of lamb, a juicy, delicately flavored dish if ever there was one. Crown roasts, rack of lamb, shanks, shoulder roasts, ground lamb, etc., all can be cooked with care and imagination to produce a meal that's a treat to eat.

Besides this poor consumption rate, which could change as more people eat good, well-prepared lamb, other problems plague the sheep industry. Predators are a major problem, both to the owners of large flocks on the western ranges, where coyotes are the prime menace, and to the farm flock owners in the East, where the worst predator is the domestic dog—people's lovable little Spots and Rovers. The predator problem can be so bad as to preclude sheep keeping altogether or to make it extremely costly in terms of fencing, where fencing is feasible in the first place. Then, too, the great upsurge in the use of synthetics affected the sheep business by affecting the wool market. Nowadays, most flocks are kept for both meat and wool, with few flocks devoted exclusively to wool production.

Sheep, however, are extremely economical little beasts, utilizing ranges and pastures more efficiently than cattle—that is, you can get more pounds of meat per acre of grassland from sheep than you can from beef cattle. Also, the animals are small and easy to handle, although good management, especially at lambing time, is critical to successful sheep farming.

Where does this leave the small producer? If you have rough, hilly land, sheep can make excellent use of it. Pastures, lawns, and meadows can all be well used by sheep, and the amount of land needed can be as little as your backyard, depending on what type of sheep raising you're doing. Keeping a small flock or raising feeder lambs requires a relatively small investment in buildings and equipment. Also, children can easily be involved in sheep-raising projects because the animals are so small and easily handled. (Being stepped on by a sheep just isn't the same as being trod on by a cow, no matter how innocent both animals may have been.) The sale of the wool or the hides, plus production of lamb for home use and for sale, can provide some income. Also, because of the scarcity and expense of lamb in the supermarket, there is quite a potential for a lamb producer to build up an excellent freezer trade business; with an abundance of grassland, this can be quite a good route to follow. All in all, if predators and fencing can be dealt with, sheep can be one of the best small-scale livestock enterprises. If you don't want an enterprise but just want to raise some meat for your own use, again, sheep can be a much simpler project than beef or pigs; and raising a few orphans or a couple of feeders can easily be added to any existing livestock projects you might have. I can't think of anything simpler than raising a few feeder pigs and lambs over the summer, never having to deal with winter weather and frozen water, and ending up with the most sumptuous meat ever.

Types of Sheep Production and Breeds

As in beef and swine production, in sheep farming also there are operations considered purebred and those considered commercial. Purebred breeders maintain flocks with good (it is hoped) backgrounds and pedigrees in meat and/or wool production. Although meat and wool are products of purebred operations, the sale of breeding stock is the major part of the enterprise. Because the sale and improvement of genetically superior animals is the "raison d'etre" of purebred operations, excellent record keeping, management, breeding, and culling practices, as well as promotional activities, are required.

Commercial operations generally use grade or crossbred animals; a common situation is one in which a superior, purebred ram of a good meat breed is used on crossbred ewes of both good meat and good wool type. Purebred ewes, in addition to purebred rams, may also be kept in commercial operations and used as breeding stock. The types of commercial enterprises vary, although the biggest distinction is between farm flocks (thirty to fifty ewes, often a sideline to some other business) and the western range sheep operations (where flocks consist of up to several hundred ewes). There also are feeder lamb enterprises, in which large numbers of weaned feeder lambs are bought and fattened for market. Feeders are usually purchased at about sixty to seventy pounds, are fed to gain one-third to one-half a pound per day, and are then sold as fat lambs at about one hundred pounds.

For the small-scale stockkeeper, obviously the farm flock is the type of operation of interest. Farm flocks may produce spring lambs, which are lambs born late in the winter and sold early in the spring at about eighty pounds; late lambs, where lambs are born in the spring and sold in the fall at about eighty to one hundred pounds; or hothouse lambs, which are specialty market lambs sold at six to ten weeks of age, weighing thirty to forty-five pounds. For the small flock owner, late lambs are most common, since lambs born during April and May can make maximum use of pasture, and building and equipment costs are kept down when the ewes lamb in milder weather. Most late lambs can be finished on forages and be ready for slaughter in October.

Most sheep breeds are classed as one of two types: fine wool or mutton; the mutton types are further divided into medium- and long-wool breeds. Fine-wool breeds are more angular bodied than mutton types and do not have particularly good mutton qualities; mutton breeds are squarer, stockier, and more compact, bred for meat but still producing a good fleece. Fine-wool breeds include the Merino, Rambouillet, and Debouillet. Medium-wool breeds are the

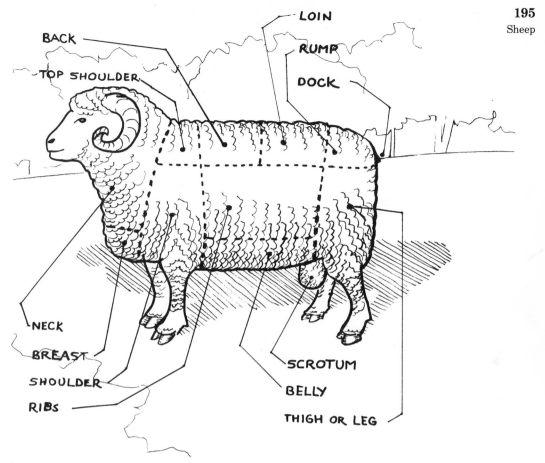

BACK

TOP SHOULDER

LOIN

RUMP

DOCK

NECK

BREAST

SHOULDER

RIBS

SCROTUM

BELLY

THIGH OR LEG

Fig. 8-1. Parts of a Ram

Cheviot, Dorset, Hampshire, Oxford, Suffolk, Shropshire, South-down, Columbia, Montdale, Corriedale, and Targhee. Cotswold, Lincoln, Leicester, and Romney represent the long-wool breeds. Karakul sheep are classed as yet another type—fur—and are the sheep used in the production of Persian lamb, broadtail, and chamois. Black-faced Highland sheep are classed as a carpet-wool type. Table 8-1 shows some characteristics of the more popular breeds.

Starting a Flock

Before getting into the ewe–lamb business, even on the small scale of just a few head, some thought should be given to the nature of sheep and the special type of care and handling they need. Perhaps more than with any other type of livestock, careful handling is essential. With virtually no defenses, sheep are timid and easily frightened; they must be handled quietly and gently at all times.

Table 8-1. Some Characteristics of Popular Sheep Breeds

BREED	CLASSIFICATION	SIZE	COLOR	HORNS	FLEECE WEIGHT
Rambouillet	fine wool	large	white	rams only	heavy
Merino		medium	white	rams only	heavy
Suffolk	meat, medium wool	large	black face and legs	polled	light
Hampshire		large	black face and legs	polled	medium
Dorset		medium	white	rams and ewes, polled or horned	medium
Cheviot		medium–small	white	polled	medium
Southdown		small	brown face and legs	polled	light
Shropshire		medium	black face and legs	polled	medium
Tunis		medium	red or tan face	polled	medium
Corriedale	dual purpose	medium–large	white	polled	heavy
Columbia		large	white	polled	heavy
Targhee		medium–large	white	polled	heavy
Montdale		medium–large	white	polled	medium
Romney	long wool	medium–large	white	polled	heavy
Lincoln	long wool	large	white	polled	heavy
Karakul	fur	large	black, brown	rams only	medium
Scottish Blackface	carpet wool	medium–large	black face and legs	rams and ewes	heavy

For this reason, too, sheep are easy prey to predators, and the small-flock owner must be especially on guard to protect the sheep from dogs. Although sheep will eat a variety of feeds, availability of good quality roughage is important for maintaining a healthy and productive flock. With proper feeding and management, disease should not be a problem, but many people think that sheep tend to give up the ship easily. "A sick sheep is a dead sheep" is a common saying, although the truth of it is probably somewhat mitigated by the experience of the shepherd. In all, sheep probably take less total labor than other stock, but at certain specific peak labor times the quality of your management will make or break you.

Then, think about the conditions that exist on your farm. How much pasture is available? What quality is it? Low lying, wet pastures are entirely unsuitable for sheep and should not be used because of the probability of severe parasite problems. Grass or grass–legume pastures are excellent, and sheep, because they will eat and thoroughly grind many common weeds, can actually improve your pastures. Road edges, lawns, and fence rows can also supply grazing areas for sheep. With *good* pasture, you can expect an acre to support as many as five to six ewes and their lambs, whereas poor pasture can support only one or two ewes per acre. Try to assess the quality of your grazing accurately; then underestimate, rather than overestimate, the possible stocking rate. Consider the availability of winter roughage as well. A ewe will eat approximately 600 to 700 pounds of good-quality legume hay over an average northern winter. Do you have hay available and the space to store it?

Next, consider breed. The various breeds of sheep have been developed to adapt to certain environmental conditions. The fine-wool breeds are noted for their gregariousness (flocking instinct), with Merinos and Rambouillets heading the list and Black-faced Highland sheep and Cheviots at the bottom, although these breeds are extremely hardy. Fine-wool types are also noted for their ability to travel long distances and for hardiness, making them especially suited to hot, desert areas where vegetation is sparse. Medium-wool mutton breeds are well adapted to areas of good vegetation and cooler temperatures, and they also tend to be more prolific. Dorsets, Suffolks, and Hampshires are noted for their heavy milk production, and Columbias and Targhees, frequently crossed with Suffolk or Hampshire rams, are known for a good carcass and a heavy fleece. Merinos, Rambouillets, Dorsets, and Tunis sheep, and crossbreds using these breeds, may be bred out of season; Dorsets are widely used to produce fall lambs. By writing to the various breed associations you can acquire a wealth of literature extolling the virtues of each breed, but for someone just starting out, availability of ewes will be a major determining factor in the type of sheep you finally get. Find out what breeds are popular in your area and why (consult

Polled Dorset yearling ewe.
Courtesy: Continental Dorset Club.

Hampshire ewe.
Courtesy: American Hampshire Sheep Association.

your local cooperative extension office) and see how they fit in with your needs.

The best time to buy ewes is in the late summer or early fall, which is usually a time when ewes are plentiful and prices low. Local ewes may be bought, or you may want to buy yearling ewes brought in off the western ranges through commission firms, dealers, or western producers. Western ewes have the advantage of having fewer parasite problems and of being available in quantity. If you want a large number of ewes of uniform type, this is the way to do it; but for a small flock, local sources will probably be your best supply of animals.

For the novice just starting a small flock, more mature ewes, four to six years old, may be more suitable than younger ones. They'll be cheaper than younger ewes, yet with proper care and attention they can raise excellent lamb crops for years to come. (Ewes may produce lambs until they are eight to ten years old.)

When selecting ewes, avoid long necks, narrow chests, shallow bodies, overlong or crooked legs, weak backs or peaked rumps. Especially when buying older ewes, look out for broken mouths with worn or missing teeth (a broken-mouthed ewe cannot eat well) and poor udders. Inspect the udders of all ewes for soundness; freedom from cuts, lumps, and bulges; and teat obstructions. Stay away from lame or limping ewes; they may have foot rot, which can spread through the entire herd, and they won't be able to forage well. Ewes should be neither fat (they may have breeding problems) nor thin (they may be carrying a heavy parasite load). Take a look at the ewe's lambs and judge them against the others in the flock; this is certainly a good indication of the producing ability of the ewe.

If you are buying purebred ewes or selecting a ram, buy only

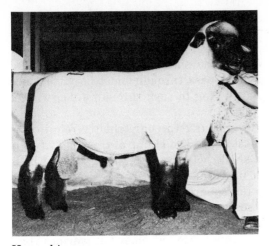

Hampshire ram.
Courtesy: American Hampshire Sheep Association.

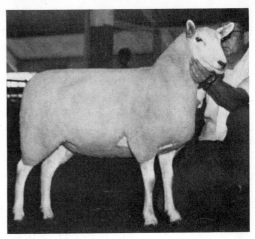

Montdale ewe.
Courtesy: Montdale Sheep Breeders Association, Inc.

from a respected, reputable dealer. In most cases, a purebred meat-type ram (such as a Hampshire or a Suffolk) of the best quality you can afford should be used on your ewes. With a small flock, you'll probably be using only one ram, so that one ram will affect your entire lamb crop. A poor ewe will transmit her weaknesses to only one or two lambs a season, but a poor ram will transmit his genes to every single lamb—which is why your ram is considered half the flock. Select him carefully, and don't settle for a bargain ram—it just doesn't pay. If you have enough ewes to justify the purchase of your own ram, try to find one who is eighteen months to three years old, sound, and fertile. If you have just a few ewes, there are still sheep farmers who will rent out a ram, with breeding charges of about five to ten dollars per head.

Managing a Ewe-Lamb Flock

Housing and Equipment

With the warmest and snuggest of wool coats wrapped tightly around them, sheep need little protection from the cold; but they *do* need shelter from wind and rain, particularly in the winter and spring. A deep, three-sided, dry and draft-free shed with a well-drained dirt floor and southern exposure is fine. Somewhat more sheltered lambing quarters may be needed where winters are severe and ewes lamb early, but don't make the mistake of confining your ewes to a warm barn all winter. (If too warm, sheep can sweat badly, then chill, and catch cold.) Indoor quarters should be well ventilated and dry and have plenty of light and sun if possible. Allow at least twenty-five square feet of shed space per ewe and lamb (twelve to eighteen square feet per ewe); a flock of ten to fifteen ewes with lambs, therefore, would need a shed of about twenty by twenty feet. Within the sheep area, all doorways and walkways that the sheep must pass through must be wide (once you've seen twenty sheep all trying to rush through a doorway at the same time, you'll see why).

During the summer on pasture, the only shelter required is shade. Obviously the simplest shade is a clump of trees, but if your fields are open, artificial shade will be necessary. Shade sheds should be at least five feet high and provide an area of six square feet per lamb and eight to ten square feet per ewe.

Lambing pens have somewhat more exacting requirements than the ewe shelters. First and foremost, they must be dry and draft-free. The lambing area must also be somewhat warmer than the ewe shed needs to be, and methods of providing supplemental heat to the lambs, which may be necessary, should be available. Temporary lambing pens, with one for every four or five ewes, can be devised as illustrated (Fig. 8-2). Lambing pens should be at least twenty square feet.

Feeders and waterers for sheep can be easily homemade. Hay-racks and grain bunks should provide twelve to eighteen inches of space per ewe (depending on the size of the ewes) and six inches per lamb. Two types of hayracks, suitable for ten to twenty ewes, are illustrated (Fig. 8-3). (Care should be taken when designing racks and feeding hay to keep the chaff off the necks and shoulders of the sheep as much as possible, since it lowers the value of the fleece.) Grain bunks should be flat bottomed, fourteen inches high,

Fig. 8-2. A temporary lambing pen can be set up in the corner of the barn by using a hinged panel. Each section should be 3 feet to 4 feet high and 5 feet long. A 5-inch strap hinge can hold the halves together. Spaces between the boards should not exceed 4 inches.

two feet wide, and four to six inches deep for ewes, and eight inches high, twelve inches wide, and four inches deep for lambs. Hay and grain bunks for lambs should be constructed inside a creep, which allows the lambs to feed but keeps the ewes out. Locate the creep in a popular gathering spot for the lambs and ewes, with the openings to the creep about eight inches wide and eighteen inches high, as illustrated (Fig. 8-4). (A particularly thin ewe may be able to squeeze into the creep, but the theory is, if she can get into the creep, she needs the feed she can steal.) If you can use the corner of the barn or shed for the creep area, you'll only need to construct two walls, making the job that much simpler. Lambs are marvelously playful and inquisitive creatures, and there's nothing they like better than climbing. If there's a convenient grain bunk for them to climb into, they will; then, of course, they'll refuse to eat the soiled feed. Partitions, similar to the hayrack design, may be needed on the grain bunker to allow only little lamb heads, not little lamb bodies, access to the grain.

Waterers may simply be tubs or pails able to be cleaned and to supply the necessary amount of water needed by the flock. Allow one inch of trough waterer space per ewe. If using pails, tie them or secure them in some way to prevent tipping. A ewe will drink from one to four quarts of water per day.

Some equipment also is needed for various management practices. Drenching for internal parasites is routine, so a drenching syringe will be needed (or a plastic bulb baster). Hand shears, pruning shears, a hoof trimming knife, and a docking and castrating tool will also be needed.

Fig. 8-3. A five-sided hay rack for ten sheep, with each side about 2 feet long, divided down the center. Openings should be 9 to 12 inches wide, depending on the size of the sheep. This rack is also good for goats. A rectangular rack, 8 to 12 feet long, can serve twenty sheep.

Fig. 8-4. A Lamb Creep. Openings for the lambs should be 8 inches
wide and 18 inches high. The boards used to reduce the
size of the opening can be removed to allow access by the
ewes if desired.

Fencing for sheep can be a problem. Actually, it's not the
fencing *for* sheep that's a problem as much as the fencing *against*
dogs. Unfortunately, the best sheep fence is expensive: woven wire,
4½ to 5 feet high. Electric fence can be used satisfactorily, but
the sheep must be trained to it and the strands spaced to be anti-
dog. The best time to teach a sheep about electric fence is either
while it's still a lamb or just after it's been shorn; the idea is to
give the lesson when the sheep does not have its fleece to act as
insulation against the shock. If the initial lesson is a good one, the
sheep will respect the fence thereafter. Two strands will effectively
keep in sheep, but three strands may be needed to keep out dogs.

Pasture

Good pasture is the backbone to any small-scale livestock operation involving ruminants, and sheep are no exception. On the
contrary, pasture is the best and cheapest source of feed for sheep,
so every opportunity should be taken to make the most of it. In fact,
lambs may be raised to slaughter weight almost entirely on pasture,
if it is of good quality.

Bluegrass, mixed with clover, fescue, or orchardgrass, makes
excellent sheep pasture. Ladino clover or alfalfa with bromegrass,
red clover, timothy, and birdsfoot trefoil make excellent combina-

tion hay and pasture crops. One acre of good pasture will carry five to six ewes; poor pasture can only carry one or two ewes.

Pasture should be made available as early in the spring as possible; fall-sown rye or wheat can often be used for early spring pasture for sheep. However, beware of turning your flock out too quickly and relying too heavily, too soon on early season pasture. Start the flock on pasture gradually in order to avoid digestive troubles. At first, turn the sheep out only for a short period, after they've been fed some hay. The grazing period can be extended and the prefeeding dispensed with over a couple of weeks. Other than digestive problems, early spring pastures can also pose nutritional ones. Early in the spring, gorgeous though it may look, the grass is still somewhat sparse and is very watery; the sheep simply cannot eat enough of it to support themselves, so some supplemental feeding will still be necessary.

A certain amount of management will be required to make maximum use of the pastures. Rotation, mowing, top dressing, and the planting of supplemental pastures may all be beneficial.

It's really a good idea to divide any large pastures you may have into two or more smaller areas to be grazed in rotation. Allowing sheep to graze one patch for two to three weeks, then moving them to a new section, will not only utilize the pasture grass to its fullest but will also greatly help in parasite control. Mowing the grass before it goes to seed also improves the grazing; sheep prefer short, actively growing plants and will skip over tall, dry, unmown grass. Top dressing pastures with manure (well rotted, if it's sheep manure) or lime will also increase pasture yields. The pasture season may be extended by the use of supplementary pastures planted to provide forage late in the season when the permanent pastures are short; second growth of hay crops is good for this purpose, as are plantings of sudangrass, rape, and kale. (Rape is best fed to lambs in combination with grass—either sow rape and grasses together, each in part of the field, or provide a little hay when lambs are grazing on all rape pasture.)

Water, shade, and salt must be provided on pasture. Trace mineralized salt should be located in the shade in containers to keep out rain and keep off lambs.

Feeding Pregnant Ewes

Because ewes are normally bred in the fall, pregnancy feeding is usually based on stored roughages used over the winter. In hothouse lamb production, however, in which ewes may be bred at any time during the year, this is not necessarily the situation. At the beginning of the gestation period, any ewes appearing particularly thin should be separated from the rest of the flock for supplementary feeding.

Ewes should gain approximately thirty pounds over the gestation period in order to be in good condition at lambing time. Occasionally ewes in excellent condition late in the fall can make the necessary gains on excellent roughage alone; but more often, roughage plus some concentrate will be required, particularly in the last one to two months before lambing. The exact amount of grain to be fed, and for how long, depends on the condition of the ewes; check them often by running your hand over their toplines.

Early in gestation, a ewe should be fed four and one-half pounds per day of good legume hay, or she can be fed legume hay or pasture ad lib. If a quantity of legume hay is not available, two pounds of grass hay plus three pounds of legume hay can be used. (A breeding ewe will consume approximately 600 to 700 pounds of good legume hay over the winter.) Corn or grass silage can be substituted for the hay at a ratio of two and one-half pounds of silage to one pound of hay. If feeding a ration high in corn silage, add one-half ounce of ground limestone per ewe per day, plus one-quarter pound of a protein supplement such as linseed, cottonseed, or soybean meal. It is better to feed a combination of silage and hay if possible, feeding two or three pounds of silage along with hay fed ad lib. Be extra careful to avoid feeding moldy or spoiled silage to sheep. Four or five pounds of roots or cabbage can be substituted for the silage if all moldy or decayed parts have been removed. (If the cabbage has been dusted for cabbage worms, remove the outer leaves before feeding it to the ewes.) If neither legume hay nor silage is available, but only poor-quality hay, feed the ewes one-quarter pound of grain, such as 14 percent dairy mix, per day in addition to the hay fed ad lib. Sheep can also make excellent use of any odds and ends of roughage that you may have. Cull beans, unsalable cabbage, bean straw or bean pods, oat or wheat straw, corn stover, corn fodder, etc. can be fed in conjunction with legume hay, plus, if necessary, a little grain.

During the last four to eight weeks of gestation, grain should be added to the ration at a rate of one-half to three-quarters pound per day. The grain can be a commercial 12 or 14 percent dairy feed, or you can mix your own feed if you have homegrown grains available. Without legume roughage, use a 16 to 18 percent dairy feed, or make up a grain mix containing a protein supplement and feed it at a rate of one to one and one-quarter pounds per day. Twenty pounds of corn, twenty pounds of oats, and six pounds of linseed meal is a good grain mix for pregnant ewes.

Fresh water and trace mineralized salt should be kept available to the ewes at all times. If nonlegume roughage is being fed, a calcium supplement, such as ground limestone, should also be supplied. One method of feeding the calcium supplement, other than just adding it to the grain, is to allow ewes free access to a mix of equal parts salt and ground limestone.

During the barn-feeding period it is important that the ewes get daily exercise. Mutton-type ewes especially can get too fat and produce weak lambs if confined to a barn and fed too well during pregnancy. If the sheep just hang around when you turn them out, try scattering some hay at a distance to encourage the ewes to walk. Exercise for fine-wool types apparently is not as important as for the mutton breeds; for them, confinement and liberal feeding does not seem to produce the same ill effects.

Just before lambing, ewes should be well fed and not forced to eat very coarse or poor-quality hay or roughage. Poor-quality roughage, very bulky rations, or a reduction in grain feeding within a few weeks of lambing can produce "before lambing paralysis," or pregnancy disease (see "Diseases"). Make the prelambing ration slightly laxative by the addition of roots or some wheat bran or linseed meal.

Lambing

Lambing time is the time of most excitement and most work around the sheep shed. Everything must be gotten ready—pens cleaned, equipment in order—in anticipation. After all, this is it—this is why you've been caring for the ewes so carefully all winter long and this is what you're willing to stay up all night for. Don't let things get slipshod now; be aware of the approaching lambing time of each ewe, get her ready, and be on hand to help out should problems arise.

At least four weeks before lambing, ewes should either be shorn or cleaned up. If they're not shorn, trim the wool from their docks; remove any loose, dirty wool tags from their sides and rear; and clip the wool around the udder.

As lambing time approaches, the ewe will develop a sunken appearance on each side of her rump, the vulva will be swollen, the udder will fill, and the teats will become swollen. Just before lambing, she may be restless, get up and lie down frequently, and paw at her bedding. An experienced ewe can be moved to the lambing pen just before lambing, but a novice had better be moved to the pen a few days before she's due so she can relax in her new surroundings. The lambing pen should be well bedded with fresh, clean straw.

Most ewes have little trouble lambing; the ewe will strain, the water bag will break, and the lamb will appear, front feet first with its head between them. If a ewe strains for more than half an hour with no result, help will be necessary. If at all possible, get an experienced person to help, or if necessary, call your vet; once you've assisted someone you will be able to do it yourself. If you must do it yourself, take every care to avoid hurting the ewe. Trim your fingernails and wash your arm thoroughly with a mild soap and water. Then lay the ewe carefully on her right side by holding her under the jaw with the left hand, reaching under her with the right hand, and grasping the right hind leg well down toward the hoof. Pull gently on the leg and lower the ewe to her side. Then lubricate your hand and arm with mineral oil or vaseline. If there is only one lamb, recognizing the presentation problem and correcting it is not usually too difficult (see Chapter 3). However, if there is more than one lamb, proceed slowly and cautiously; not only is there much less room to work in, but you must be careful to sort out which legs go with which lamb. As with other livestock, a posterior presentation (both hind legs first) can be assisted without going inside the ewe, and it should be helped immediately. As the ewe pushes, pull the lamb down and out in the direction of the ewe's hocks. Speed is of the essence to prevent suffocation of the lamb, but still, pull only *with* the contraction.

After one lamb has been born, if there is any doubt about whether there might be more, it may be worthwhile to check the inside of the ewe as described above.

Ewe and her lambs.
Credit: USDA photo.

As soon as the lamb is out, if it appears to be breathing normally, well and good. Dip its navel in iodine to prevent navel ill, and see that the lamb nurses within thirty minutes; most lambs are up and nursing quite soon after birth. Make sure the lamb can be kept dry and free from chills. At this point you should check the udder of the ewe to be sure that both teats are functioning and that there is no wool interfering with nursing. Actually, the udder should be checked daily for several days to be sure the lamb is taking all the milk; if the ewe is producing more than the lamb will drink, milk a little out of the udder until the lamb is large enough to consume the full amount.

Problems may arise if the lamb appears weak or chilled, or if the ewe rejects her lamb. Should a lamb appear lifeless or weak at birth, *immediately* wipe the mucous from its mouth and nose, and then slap it on both sides of the body just behind the shoulders, or rub it vigorously with dry straw or a burlap bag. You may have to help the lamb nurse at first by holding it up to the udder of the ewe and forcing a little milk into its mouth. If this step doesn't work, the ewe may have to be laid on her side and rump to make it easier for the lamb to find the teat. If the lamb absolutely won't nurse, milk out the ewe a little, feed the lamb with an eyedropper or teaspoon, and try again later. Eventually, the lamb will probably nurse; it will just take some perseverance on your part.

If the lamb appears chilled, warm it up as quickly as you can: wrap it in a woolen cloth and put it in a box with a heat lamp, hot-water bottle, or warmed bricks; or bring the lamb into a warmed room; or immerse the lamb, except the head and neck, in a bucket of warm water (as hot as you can tolerate on your elbow) and then dry it thoroughly. Feed the lamb after it has been revived, and reunite it with the ewe as soon as possible.

Usually, rejection of the lamb by the ewe is not a problem when the lamb and ewe are kept together in the lambing pen for a day or two. At first, the ewe recognizes her lamb by smell, so if the little lamb wanders off, the ewe might not be able to find it again. After a few days, though, the ewe can also recognize the lamb by sight, so keeping them penned together for the first couple of days can reduce the problem of disowned lambs. (Rubbing some milk on the nose of the ewe and on the rump of the lamb is also said to help.) Usually, if ewes are in good condition and have a good milk supply, they will make good mothers; thin, undernourished ewes are the poor mothers, who will often abandon their lambs. In large flocks, it may pay to mark each ewe and her lamb with a painted number (using a paint that will scour out of the wool) to enable you to immediately spot any lamb who may be lost or poorly cared for.

Occasionally, a ewe may die during lambing, or she may not have enough milk for one reason or another, or she may have too many lambs for her milk supply to handle (like triplets); in these situations, one would have orphan lambs to deal with. In some cases, especially in large flocks, there will often be a ewe who has lost her lamb at the same time that there's an orphan to be adopted—a matched set, if you can just get the ewe to accept the little stranger. The classic adoption technique is to tie the pelt of the ewe's dead lamb onto the lamb to be adopted, removing the pelt after a few days. Another method is to restrain the ewe in a stanchion or to tie her with a halter so that the lamb can nurse, but the ewe must be allowed to exercise a few times a day. It should not be more than a few days before she'll accept the lamb. Yet another trick is to put the ewe and lamb in a pen and tie a dog up nearby, thus creating a burst of maternal protectiveness, resulting in an instant bond. It is possible that it will take as much as a week for the adoption to fully take hold with any of these methods, or it may never happen at all, but patience usually wins out in the end. However, it is certainly worth the effort on your part to see that it does happen; the alternative of hand raising the orphans is a lot more work (see "Raising Orphans").

Feeding Ewes during Lactation

Immediately after lambing, the ewe can be fed all the hay she'll eat. Grain, however, should be withheld for the first twelve hours, then fed in increasing amounts, allowing a handful more each morning and evening until the full feed level is reached after about ten days. Silage and root crops, which tend to increase milk flow, should not be fed for the first several days postpartum in order to reduce the chances of swollen udders.

At this point, the best hay possible should be fed to the ewes; good legume hay will supply some of the extra protein needed by the lactating ewe. In addition to hay fed ad lib, she should receive a 14 percent grain mix, the quantity fed depending on the individual sheep. Large, thin ewes nursing month-old twins may need as much as two pounds of concentrates daily, whereas fatter ewes nursing single lambs may need as little as three-quarters of a pound. If legume hay is not available, increase the protein content of the grain mix.

Of course, water, salt, and minerals must be available at all times.

Care of Young Lambs

Creep Feeding If lambs are born more than a month before the start of the pasture season, a creep-feeding area should be set up as previously described by the time the lambs are five to ten days old. Excellent legume hay should be supplied in the creep daily; lambs won't eat picked over hay, so the rejected hay can be fed to the ewes or other stock. The same is true for the grain: it must be supplied fresh daily, as much as the lambs will eat, and leftovers fed to the ewes. Twenty pounds of cracked corn, twenty pounds of whole or crushed oats, twenty pounds of wheat bran, and ten pounds of linseed meal can be used as a creep feed, as can a 16 percent dairy mix. Usually lambs will start eating grain readily, particularly a

Table 8-2. Concentrate Mixtures for Sheep and Lambs

A. For ewes fed good legume hay for roughage:
    1. Corn, 40 lbs.; oats, 30 lbs.; wheat bran, 20 lbs.; linseed meal, 10 lbs.
    2. Oats, 60 lbs.; corn, 25 lbs.; wheat bran, 15 lbs.
    3. Corn, 60 lbs.; wheat bran, 40 lbs.
    4. Corn, 90 lbs.; linseed meal, 10 lbs.

B. For ewes fed good legume hay and silage* or roots:
    1. Oats or barley, 66 lbs.; wheat bran, 34 lbs.
    2. Oats, 45 lbs.; corn, 40 lbs.; wheat bran, 10 lbs.; linseed meal, 5 lbs.
    3. Oats, 45 lbs.; corn, 45 lbs.; linseed meal, 10 lbs.
    4. Corn, 60 lbs.; wheat bran, 30 lbs.; linseed meal, 10 lbs.

C. For ewes getting early-cut hay containing about 50 percent legumes and with or without a moderate amount of silage* or roots:
    1. Oats, 20 lbs.; corn, 50 lbs.; wheat bran, 20 lbs.; linseed meal, 10 lbs.
    2. Corn, 60 lbs.; wheat bran, 25 lbs.; linseed meal, 15 lbs.
    3. Oats, 50 lbs.; barley, 40 lbs.; linseed meal, 10 lbs.
    4. Barley, 80 lbs.; wheat bran, 10 lbs.; linseed meal, 10 lbs.
    5. Oats, 50 lbs.; corn, 20 lbs.; wheat bran, 20 lbs.; linseed meal, 10 lbs.

D. For ewes receiving early-cut hay containing only a small percentage of legume:
    1. Oats, 40 lbs.; corn, 30 lbs.; wheat bran, 20 lbs.; linseed meal, 10 lbs.
    2. Barley, 65 lbs.; wheat bran, 25 lbs.; linseed meal, 10 lbs.
    3. Oats, 50 lbs.; barley, 40 lbs.; linseed meal, 10 lbs.

E. Creep feeds for lambs in the barn but also with access to legume hay or for creeps on pasture:
    1. Oats, 20 lbs.; corn, 20 lbs.; wheat bran, 20 lbs.; linseed meal, 10 lbs.
    2. Barley or corn, 30 lbs.; oats, 20 lbs.; linseed meal, 10 lbs.
    3. Oats, 75 lbs.; wheat bran, 15 lbs.; linseed meal, 10 lbs.
    4. Oats, 20 lbs.; corn, 30 lbs.; wheat bran, 10 lbs.; linseed meal, 10 lbs.; dehydrated alfalfa, 10 lbs.
    5. Commercial calf starters are also excellent creep feeds for lambs but are more expensive than the above-mentioned rations.

F. For lambs after weaning, intended for market, which have access to good pasture or good legume hay:
    1. Corn, 90 lbs.; linseed meal, or soybean oil meal, 10 lbs.
    2. Corn, 80 lbs.; dried brewers' grains or distillers' corn dried grains, 20 lbs.

    In the foregoing mixtures, cottonseed meal, or soybean oil meal, may be substituted for the linseed meal. When the price is favorable, such feeds as dried brewers' grains, distillers' corn dried grains, and corn gluten meal may be used as a source of protein.

*Corn silage or silages made from grass or pea-vines may be fed.
Source: J. P. Willman, W. F. Brannon, D. E. Hogue, Cornell Extension Bulletin 828.

sweet feed, but they may occasionally need some encouragement. Try sweet feed if you're not already using it, or try to get the lambs to eat some grain out of your hand; once one lamb eats the grain, they all will, so give them free access to the creep at all times. By the time they are about two months old they will be eating about half a pound of grain each per day.

If the lambs are born close to the time they're to go out on pasture—assuming it's good pasture—it's not really necessary to use creep feeding. However, if your pastures are not that good, if the ewes seem a bit thin, or if several ewes are nursing twins, creep feeding is a good idea. Continue the lambs on creep feeding during the pasture season until weaning time if you are interested in quick gains. Occasionally, though, lambs won't eat grain out of a creep on pasture, even if they had learned to eat from a creep in the barn. Adjusting the creep so the ewes can go in as well as the lambs will usually get the lambs in and eating alongside their moms. Once the lambs use the creep readily, it can be readjusted to exclude the ewes—a dirty trick, but it works. Of course, once the gang is on pasture, there is no need to provide hay in the creeps.

Docking Docking, or removing the tail, is done when the lambs are from three days to three weeks old, the younger the better. Lambs are docked to keep them clean and to facilitate the breeding of ewes. If you've ever seen a long-tailed sheep, you know that the tail is virtually useless and just manages to collect dirt and droppings; it is doing no disservice to the lamb to remove its tail, which may be done in a few ways. No matter which method is used, however, the lambs to be docked should be put in a clean, freshly bedded pen; afterwards, they can be returned to the ewes. One technique is not very complicated: cutting off the tail with a disinfected sharp knife, hatchet, etc. One person holds the lamb up on its rump, set against something clean and hard, and the other then cuts the tail off one-half to one inch from the body. As long as the operation is performed while the lamb is small, bleeding will not be a problem. Another method is to use an elastrator, a tool used to place a very tight rubber band around the tail one-half to one inch from the body. The band cuts off the blood supply to the tail, which eventually drops off. This method should not be used during fly season, however.

Castration Ram lambs are usually castrated, often at the same time they are docked, but before they are three weeks old. With either docking or castrating it's best to learn how to do it by assisting an experienced person first. Castration of lambs is just like castration of calves. Have the lamb held on its rump at a comfortable working height during the operation (see Chapter 6 for castration techniques). With both docking and castration, observe the lambs closely for several days afterwards for signs of infection.

When raising lamb for your own use, it is not really necessary

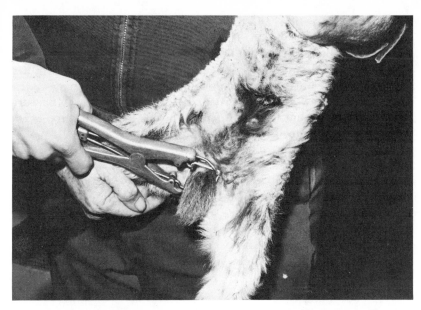

A lamb being castrated with an elastrator. A rubber band, placed
around the testicles close to the body, cuts off circulation, causing
the testicles to wither and drop off.
Credit: USDA photo, M. Lemmon.

to castrate the ram lambs if they are to be slaughtered at around
five or six months. Unlike boar, as compared to barrow meat, meat
from ram lambs tastes no different than meat from wethers (castrated lambs). However, wethers are said to produce a better carcass, and ram lambs are discriminated against in the marketplace.

Weaning

Most lambs are weaned at around four and a half months, but
some may be weaned earlier (three and one-half months) or later
(five months), weighing from sixty to one hundred pounds. If the
ewes are milking well and the pastures are good, there is no advantage to early weaning. If conditions are such, however, that the
lambs and ewes would do better separately, then the lambs should
be weaned early.

At weaning, the best pastures possible should be made available to the lambs, and the lambs and ewes should be separated
permanently. Feeding the lambs one-half to three-fourths of a
pound of grain while on pasture keeps them gaining well. If the
lambs are weaned early and put on full feed, vaccination for overeating disease, which occurs with heavy grain feeding, is recommended.

While the lambs are put on pasture at weaning time, the ewes
should be gathered into the barn for observation. Cut out the grain

portion of their ration, and reduce the hay intake for a few days to encourage them to dry up. If a ewe is milking very heavily, she should be milked a little a day or so after the lamb was removed. Check her again two days later, and continue to milk her lightly every other day until there seems to be no danger of damage to the udder. After a few days, the ewes may be returned to the pasture (the second best), but they still should not be fed grain until a few weeks before the breeding season.

At this point, you may want to select your replacement ewe lambs. These lambs should come from the heaviest milking, heaviest shearing, and most prolific ewes in the flock. The best grown, older twin lambs are often the best choice. Generally, ewe lambs should not lamb until they are at least yearlings, but if they are particularly vigorous and well fed, they may be bred as early as eight months old. Many sheep breeders, however, delay breeding their ewe lambs to have them lamb at one and a half or two years old.

Breeding

The type of lamb production system you're following will determine the breeding season for your ewes. Most small-flock owners aim to have their ewes lamb in the spring, usually April or May, and so turn their ram in with the ewes around November or December. Hothouse lamb producers can breed at any time of the year, and for early spring lambs, breeding starts around August.

Ewes are usually seasonal breeders, coming into heat in the fall, beginning about August, and cycling every sixteen to seventeen days with a heat period of one or two days. A few weeks before breeding season, the ewes should start to gain weight. Getting them gaining before breeding increases the conception rate at the first service (which results in a more uniform lamb crop) and also tends to increase the incidence of twinning; a small flock can get as much as a 150 percent lamb crop, with good management (and good luck). This "flushing" can be accomplished by switching the ewes to better pasture, or more easily, by feeding one-fourth to two-thirds of a pound of grain per ewe daily. Adding pumpkins, cabbages, or roots to the diet will also help flush the sheep. The grain ration is continued for about a month after the ram is put in with the ewes.

The ram, too, should be in condition for breeding. He should be gaining and should have had plenty of exercise before the breeding season. Feeding a small amount of grain throughout the season to a heavily used ram will improve his performance, but an overconditioned ram will not be a good breeder. If the ram is carrying a heavy fleece as the breeding season starts, he should be shorn if the weather is still warm, or at least, some wool should be removed from his neck, breast, belly, and scrotum. Ewes should be clipped around the vulva and dock. At this time, both rams and ewes should have their feet trimmed if necessary.

In order to identify which ewes have been bred and when, a marking system is used. A marking harness on the ram will do the job, or you can paint the chest of the ram every few days. Either way, a mark is left on the rump of the ewe. The marking system works as follows: three pastes are made by mixing lubricating oil with yellow ochre, venetian red, and lamp black. The ram is started off with yellow, the paste being renewed every few days. After sixteen days, the color is switched to red; then after the next sixteen days, to black. This system shows whether the ewe was bred on her first, second, or third heat, which will tell you when to expect the ewe to lamb. If *all* the ewes have been rebred on their second heat, the ram may not be very fertile, and you had best switch to another ram.

Generally, the ram is simply allowed to run with the flock. Yearling or two-year-old rams can serve forty to fifty ewes during the season, and a well-grown, properly fed and managed ram lamb can breed twelve to fifteen. If large numbers of ewes are to be bred, if the ram is old, or if he is a lamb, turning him out with the ewes only at night may be more satisfactory; this is also a good practice during hot weather. Allowing the ram to remain with the flock for a few months after breeding is fine, as long as the ram doesn't tend to boss around and injure the ewes. Also, if you allow the ram to stay, you must be prepared for a few very late lambs.

Shearing

Shearing is often done in the spring after lambing, when cold, wet weather has passed. However, many sheep people feel that shearing the ewes *before* lambing makes them take care of their lambs better by forcing them to take shelter during inclement weather; it is recommended to shear two or three weeks before lambing, if that route is followed. In any event, most small-flock owners find it most convenient to have a custom shearer do the job.

Fleeces should be dry at shearing time and as clean as possible. A clean, dry, wooden floor or a pen spread with a heavy canvas can serve as the shearing area. Sheep waiting to be shorn should be kept where they will not get dirty or be able to track dirt, straw, etc. in with them to the shearing floor.

A good shearer takes the entire fleece off evenly and close to the skin, in one piece with no second cuts, and without injuring the sheep. (Second cuts produce short wool of lower value.) After the fleece is off, spread it out on a clean, dry floor (not dirt) with the skin (flesh) side down. Remove all dungy tags, sweat locks, and short leg wool from the fleece and put them in a burlap bag for separate sale. Fold the sides of the fleece into the center; then roll up the entire fleece, starting from the rear end. Tie it with paper wool twine—as opposed to binder twine, heavy, rough jute twine, or any other twines that would shed loose fibers into the fleece. Do not use wire. Place the tied fleece in a regular wool bag or in a clean feed bag, and store it in a clean, dry place until marketing.

Depending on the breed of sheep, the fleece may weigh anywhere from four to twelve pounds; the average in the United States is eight pounds. Consult your county agent for cooperative wool pools in your area, not only for marketing your wool but also for locating sheep supplies. Last year, fleeces were getting seventy-nine cents per pound, or about six dollars for a well-prepared, average weight fleece. As of this writing, however, sheep producers can count on a minimum of seventy-two cents per pound for wool. The USDA has an incentive payment program for wool producers which guarantees this price: you sell the fleece and the USDA makes up the difference should it have been sold for less. Consult your local Agricultural Stabilization and Conservation office for details and applications.

Routine Health Practices

Routine health care consists primarily of parasite control, both internal and external, foot care, and possibly some vaccinations.

1. *Dip, dust, or spray* at least once a year to control external parasites. Consult your local county agent for the best products to use in your area.

2. *Drench ewes and lambs* with phenothiazine in the spring before going onto pasture and again when returning to winter quarters in the fall.

3. *If pasture conditions* are conducive to heavy parasite infestations (not rotated, poorly drained), drench the flock every three to four weeks, alternating the drenching compound. Consult your county agent or vet for drenches and their use.

4. *Drench breeding stock* just before the breeding season.

5. *Examine feet* periodically and trim when necessary, or at least once a year. Set the sheep on its rump for trimming, and using a knife or pruning shears, follow the outline given in Chapter 4.

6. *Vaccinate ewes* one month to six weeks before lambing for enterotoxemia, or overeating disease, as recommended by your vet.

7. *Vaccinate lambs* for sore mouth at one month of age if it has occurred in the flock, under the recommendation of your vet.

8. *Vaccinate lambs* for overeating disease at one month if they are to be weaned early and put on full feed.

Diseases and Problems

Pregnancy disease, already mentioned, affects ewes within one month of lambing; it occurs only in ewes who have been underfed during the last month or two of gestation or those who have been fed poor roughage and little grain. It is also more common in ewes carrying twins or triplets and in older, rather than younger, ewes. Lack of exercise or overexposure to adverse weather conditions may tend to bring on the disease. The earliest symptoms are sluggishness and unsteady walking. As the disease progresses, the ewe becomes paralyzed and may lie with her head bent back. Call your vet immediately, since in its early stages the disease may be checked.

Stiff lamb disease, or white muscle disease, affects lambs from two to six weeks old. Affected lambs have difficulty walking and appear stiff or somewhat paralyzed in their legs. The lambs can nurse if they are helped to the ewe, but many stiff lambs die. Apparently, the disease is caused by a lack of Vitamin E and the mineral selenium. Preventative feeding is the best method of control. Green plants are high in Vitamin E, as is unextracted wheat germ meal. Linseed meal or wheat bran fed to the ewes for sixty to ninety days, starting thirty days before parturition, will provide adequate selenium. In the early stages, the disease may be reversed by giving sixty to one hundred milligrams of Vitamin E (alpha tocopherol) orally to the lambs for several days. Consult your vet.

Sore mouth shows itself as sores or scabs on the lips of lambs or on the udders of nursing ewes. Applying tincture of iodine or a saturated solution of potassium permanganate to the sores after the scabs have been removed may cure it. If the case is severe, consult your vet about treatment and vaccination.

Constipation in lambs may be relieved by giving one to two teaspoons of warm castor oil or an enema of warm, soapy water in conjunction with an oral dose of one-fourth to one-half ounce of mineral oil.

Pinning occurs when feces stick to the wool of the tail and "glue" the tail to the body, which would, of course, interfere with normal elimination. Lambs can die from pinning if it is not taken care of quickly. The tail must be pulled free and the entire area cleaned. A handful of dust or sand tossed over the rear end helps prevent reoccurrence.

Sore eyes occur occasionally and are evidenced by profuse tearing. The eyes can be treated twice daily with a saturated solution of boric acid, administered with an eyedropper.

Navel ill causes lameness or swelling in one or more joints of the leg. It is caused by an organism entering the body through the navel cord, and it may be prevented by dipping the cord at birth in tincture of iodine.

Parasites, both internal and external, were covered under "Routine Health Practices," as was *hoof trimming.*

Predators, next to parasites, are probably the cause of most damage to the small flock. For most small-flock owners, the predator of concern is the dog. A sheep is easily killed by a dog, not only by being directly attacked but also by being run and literally scared to death. A sheep run by a dog can easily drop, suffocate, and die, without ever having been touched by the dog. The best defense against dogs is, unfortunately, a good fence. Next, alert all your neighbors that you are keeping sheep and that, not only will they be liable for any damage done to your sheep by their dog, but also they may end up with no dog. The sheep owner is perfectly within his or her rights to shoot any dog seen on the property, regardless of whether it is actually chasing the sheep. It may seem like a terrible thing to do, but once you see a dog mangling your sheep, there will be little hesitation on your part. And there shouldn't be *any* hesitation: one dog can wipe out your entire flock in depressingly little time, so it's a choice between your flock or the dog.

Raising Feeder Lambs

What if you really like lamb but don't want to be, or can't be, involved with all the business of keeping a ewe–lamb flock? Easy! Raise a few feeder lambs or orphans. Either way, the lamb-raising process is short term and fair weather and will put excellent lamb in your freezer. Raising feeders is simpler than raising orphans, involving little time, space, or equipment. Feeder lambs are bought at a weight of 60 to 70 pounds; you must raise more than one for companionship's sake. Over a seventy-five- to ninety-day feeding period, the lambs are brought to a slaughter weight of about 110 to 115 pounds. Generally, feeder lambs are confined—not pastured— and fed large amounts of grain with little hay. Housing for the feeders need consist of only a bedded pen or a shed equipped with a hayrack, grain bucket, and water bucket. About 400 pounds of grain plus about 50 to 100 pounds of hay will be required for finishing each lamb. The feed can be either a commercial complete feed or a mix of homegrown grains; hay should be good-quality legume. Then, all you have to do is keep the animals' quarters clean and dry and feed and water the lambs. Some hints on feeding follow:

1. Bring lambs up to their full grain ration *gradually.*
2. Feed twice daily.
3. First feed the grain, wait for it to be finished, and then feed the hay.
4. Reduce the amount of grain fed to a quantity that is cleaned up before the next feeding. All the grain ought to be consumed within about fifteen minutes.

5. Provide one foot of bunk space per lamb.
6. Keep feed clean and fresh, and clean bunks daily.
7. Adjust the rate of hay feeding to the quality of the hay (feed more hay if it is poorer quality).

17

Sheep

Some suitable rations for feeders would be:

2 pounds shelled corn, barley, wheat, oats, or a mixture of these
1½ pounds good-quality legume hay
$^1/_{10}$ pound protein supplement (which is 40 to 44 percent protein); trace mineralized salt, free choice

or

2½ pounds ground ear corn
½ pound good-quality legume hay
$^1/_{10}$ pound protein supplement (40 to 44 percent); trace mineralized salt, free choice

Raising Orphans

Raising orphans is more challenging and more fun than raising feeders, and it is definitely more work. It is, however, an excellent short-term sheep project for the small-scale stockholder and a terrific way to test the waters of sheep keeping before diving into the responsibility of owning a full ewe–lamb flock.

You may have acquired an orphan lamb from your own flock, or you may be buying a couple of orphans as a small-scale sheep project. If the lamb is coming from your own flock, it will probably be a newborn, needing colostrum within an hour or so of birth. The colostrum may be some that you milked from other ewes and then froze, just in case of such an emergency, or you may mix up a colostrum substitute from twenty-four ounces of warm cow's milk, one beaten egg, one scant teaspoon of cod-liver oil, and one scant teaspoon of sugar. After the first feeding, give three to four ounces of colostrum every four hours for the first twenty-four hours.

If you are buying an orphan, or a bottle lamb, as it is often called, try to get as old a lamb as you can. Orphans come on the market anywhere from one day to a few weeks old, but the older it is when you get it, the more likely it is to live. Early spring is the usual time to find bottle lambs.

Set up a pen for the lambs before you get them. The pen should be deeply bedded, clean, and dry (dryness is *extremely* important), and if the weather is still cold, some supplemental heat may also be necessary, especially if the lambs came from a warm barn. A heat lamp suspended a couple of feet above the bedding is an excellent heat source; just be sure to hang it high enough or to put a guard around the bulb, so the lambs don't bang into it and burn their noses. You'll have to use your judgment about the amount of heat. If the lambs are all huddled under the lamp and shivering, obviously more heat is needed. If they're wandering around the pen happily and sleeping away from the lamp, you can probably turn it off. It may be that heat will be needed only during the night for a short while.

By the time the lambs are two to three weeks old, they'll start nibbling at grain and hay, so a feeder should be included in the pen. The usual pattern of a lamb when confronted with a feeder is to look at the hay and grain in it; to walk all over the feed, checking it out; and then to decide that although interesting, the stuff is just too dirty to eat. So you have to build a feeder that will keep them from doing this. Anything that works will do, including the use of small buckets for the grain (dimensions for traditional hay and grain bunks for lambs were given previously). A bucket held securely in place is a fine waterer for the lambs.

When you first get your lambs, their source of nourishment will be milk, fed by you in a bucket or from a bottle. The bucket can be of the LamBar type: a pail with a lamb nipple. Don't use a calf pail—calf nipples are shaped differently and the lambs may not take to them. For bottle-feeding, don't use plastic baby bottles, which may collapse, but rather some such type as a Coke or beer bottle fitted with a lamb nipple (purchasable from most farm supply stores).

Evaporated milk mixed half and half with water can be used for the lambs, as can raw cow's milk. Milk replacer can also be used. The best is lamb milk replacer, which is designed specifically for lambs, but calf milk replacer can also be used if it is mixed one-third richer than the directions indicate (ewe's milk contains more fat than cow's milk).

During the first week, feed the lambs four or five ounces of lukewarm milk four times a day. Last year I raised three bottle lambs on milk replacer, and to make things a little easier, I mixed up a day's batch of milk at one time and refrigerated it. Then at each feeding, I filled the bottle and held it under hot water until the chill was off. Keep the temperatures and quantities of milk consistent from feeding to feeding. Remember that there is more danger from overfeeding than from underfeeding, so go easy initially and see how the lambs handle the feedings. Should a lamb start to scour, immediately halve its milk allowance and double check the cleanliness of your feeding equipment, the temperature of the milk, and if you are feeding on a regular schedule. There is always some individual variation, though. Of the lambs I raised, two were quite vigorous and drank up their allotment eagerly and completely; the other was somewhat weaker—I had to hold her on my lap and encourage her to drink, and even then she took less than the others. All three lambs, however, ended up doing quite well.

When the lambs are two to three weeks old, they can get five to six ounces of milk three times a day. At this time, start feeding the lambs some high-protein feed. Lamb pellets are standard, or you can use sweet feed mixed with pellets (the lambs may quickly learn to polish the sweet feed off the pellets, leaving quite a neat pile of shiny little pellets) or sweet feed mixed with Calf Manna. Also provide good-quality legume hay. Feed and hay should be fed fresh daily, and the rejected leftovers from the previous day fed to other stock.

As the lambs seem able to handle it, you can increase their milk allowance to a quart per lamb per day, fed in two feedings. The idea, though, is to have the lambs eating hay and grain in quantity as soon as possible so they can be weaned from the costly milk replacer. By the time the lambs are five to eight weeks old, they should be eating solids well enough to be weaned. To wean, you can either slowly cut back on the amount of milk being fed, dilute the milk with more and more water, or just stop feeding the milk altogether. I like cutting back gradually—the expressions of outrage on their little faces are so cute as they reach the bottom of their bottles prematurely.

Once they're weaned, continue the lambs on a 14 percent protein feed plus good legume hay until the pasture season. If you have a grassy area or pasture, you can fence them in, but beware of dogs. Depending on how much attention you showered on your lambs, you may be able to just leave them out on the lawn to do the lawn mowing (under your supervision, of course) without worrying that they'll wander away. Continue to feed the lambs grain, approximately one pound per day, but eliminiate hay once they are on pasture. (Of course, if the orphans were part of your flock, they can rejoin the flock as soon as it is convenient for you to have them in with the others, usually after they are weaned.)

Housing needs at this point are minimal; some simple, dry shelter or shed is all that's needed. The lambs should have been docked by the time they were two weeks old, but there is no need to castrate the ram lambs. Shearing is also not necessary, although the lambs might make faster gains if shorn.

From this time on, the lambs need only to be fed and watered, and they should be ready for slaughter by midsummer or early fall (assuming you got them in the spring).

Now I know what some of you are thinking: how can you raise those adorable little lambs on bottles (you're their surrogate mother, for heaven's sake) and then kill them? Well, you're right—it's hard. But from day one, you have to tell yourself that these animals are destined for the freezer. Don't name them. I tried to be good about this—didn't name them at all—but every time I went to feed them or saw them, I said, "Hi, lambs" or "Come and get it, lambs." It didn't take them long to learn to respond to their collective name. It got to the point where I didn't have to fence them in;

I could be shepherd and say "Here, Lambs" and they'd come running. That was the point I realized I was in trouble. After looking through their feed bills time and again, I came to the inevitable conclusion: I had to wean myself from the Lambs. I stopped playing with them, I stopped sitting around with them, I had other people feed them, and most of all, I stopped calling them Lambs and avoided calling them anything else—I just didn't call them. Slowly and painfully it worked; they stopped responding to me any differently than they responded to anyone. But, sad as it was then, I was glad later. We kept two lambs for ourselves and sold one, and ended up with 120 pounds of the most delicious (and quite anonymous—we didn't know which lamb chop was who, after all) lamb I've ever eaten.

References

BRADBURY, M. *The Shepherds Guidebook.* Emmaus, Pa.: Rodale Press, 1977.

ENSMINGER, M. E. *Sheep and Wool Science.* Danville, Ill.: The Interstate Printers and Publishers, 1952.

JUERGENSON, E. M. *Approved Practices in Sheep Production.* Danville, Ill.: The Interstate Printers and Publishers, 1973.

NORDBY, J. E. *Selecting, Fitting, and Showing Sheep.* Danville, Ill.: The Interstate Printers and Publishers, 1962.

SHEEPMAN'S PRODUCTION HANDBOOK. Denver, Col.: Sheep Industry Development Program, Inc.

SIMMONS, P. *Raising Sheep the Modern Way.* Charlotte, Vt.: Garden Way Publishing, 1976.

IV.
POULTRY

9. CHICKENS

9
CHICKENS

It's a great scene: you're standing in front of the barn, tossing out cracked corn, calling, "Heere chick, chick . . . heere chickies . . ." and twenty crazy birds come careening around the corner, flapping, half flying, galloping as fast as their little legs will allow. They screech to a halt right by your feet and start pecking and scratching madly. It never fails to make me laugh—seems like an even better way to start the day than a bowl of corn flakes.

Chickens are among the easiest of animals to keep on a small holding, and despite their oft-maligned character, can be lots of fun. People have said that chickens are dirty, smelly, stupid, and dull. Not so. Dirty and smelly tend to go hand in hand. Any living thing that's kept in dirty conditions will probably be dirty and smelly, but I stress "dirty conditions"; no animal is, in itself, intrinsically dirty, and chickens are no exception. Give them enough space, clean their pen once in a while, and you'll have happy, clean, odorless birds. As far as dumb—well, maybe they are kind of dumb, but in a very endearing sort of way. They almost seem to have a kind of group intelligence, though; any individual chicken may be a little slow, but as a group they can learn chicken-level stuff (like how to use a door to the outside, and then come in again). To top it all off, watching flock behavior patterns is anything but dull; the only people who might think chickens are dull can only be people who've never kept any.

With low initial investment, minimal housing requirements,

and no huge commitment in time, keeping hens is a fairly easy task. Most backyard flock owners keep chickens for eggs, with meat as a by-product. Raising birds solely for meat can be a little trickier than keeping layers, since raising chicks requires somewhat more equipment and dedication.

As long as you don't go overboard on equipment and housing, keeping layers can be a source of economical, as well as high-quality, eggs. The difference between a store-bought commercial egg and a fresh egg from a free-running hen is quite visible. Frequently, commercial eggs are stored for extended periods of time, resulting in a flat, runny white, whereas a fresh egg white holds together and stands up. Hens fed commercial diets produce yellow-yolked eggs; hens with access to vegetation, earth, etc. produce yolks of a deep orange. This difference in the yolk may or may not reflect a difference in the nutritional value of the egg; but it has been shown that increased levels of vitamins and minerals in the diet of the hen, as would be the case with a bird on free range, result in higher vitamin and mineral levels in the eggs.

The economics of raising meat birds is more variable. One thing commercial chicken producers produce is relatively inexpensive meat. What they sacrifice is taste, something you're not really

A farm flock of poultry.

Partridge Cochin Bantams.
Courtesy: American Bantam Association.

aware of until you've tasted a range-raised chicken. Of course, taste is a personal question, and the taste of a chicken depends on many factors, but many people agree that home-raised chicken is a far superior product to today's commercial birds.

As you can see, there are several variations to keeping chickens, but basically, there are three systems: (1) keeping hens solely for eggs, disposing of culls; (2) raising only meat breed chicks from one day old to slaughter age; (3) raising dual-purpose breed chicks, keeping the hens as layers and using the cockerels for meat.

The type of chicken keeping in which you're interested will determine the breed of bird you choose.

Breeds

Although breeds of chickens are usually classified by place of origin (American, Asiatic, English, and Mediterranean), it is sufficient for our purposes to think in terms of heavy meat types, light egg producers, and middle ground dual-purpose birds. Then, of course, there are innumerable crosses of these types.

The classic commercial bird in U.S. layer production is the White Leghorn, a lightweight, white egg layer, and a veritable egg factory, producing 230 to 250 eggs per year.

Meat birds are often crossbreeds, but meat breeds include the Cornish, White Plymouth Rock, and the Jersey Black Giant.

Dual-purpose breeds are traditionally Rhode Island Reds, Barred Rocks, and a cross between the two, which results in black

White Crested Black Polish Bantams.
Courtesy: American Bantam Association.

hens and barred cockerels, called sex-links. All are brown egg layers. More exotic breeds also kept as dual-purpose birds include the
Brahma, Cochin, Autralorp, and Orpington. Compared to the dual-
purpose breeds, the light breeds generally lay more eggs per year
and consume less feed, but their productive period is shorter and
they make poor table birds. I've always kept the sex-linked pullets
and have been very happy with them; some of those hens kept up
excellent production for two years and were quite good-sized meals
when their days were over. Also, when allowing your hens range,
the extra feed required by the heavier breeds is largely made up by
their own scavenging.

Aseel and Malay
Bantams.
Courtesy: American
Bantam Association.

Starting a Laying Flock

Keeping a laying flock can begin in several ways. You can buy day-
old chicks straight run (both sexes), raise the cockerels to about ten
weeks for meat, and raise the hens to lay (at about twenty-two
weeks); or buy ready-to-lay (twenty-week-old) pullets; or pick up
mature, already laying hens from a poultry house that is culling
out its older birds. All have their advantages and disadvantages.

Chicks are almost irresistibly cute: fluffy, fuzzy, and fun to
watch. However, they do require a fair amount of care and attention
and won't start laying until they're approximately five months old.
Counting the value of the cockerels, raising your own layers can be
a little cheaper than buying pullets, but a lot depends on your setup
and equipment costs. When calculating the economics of raising

Black Rosecomb Bantams.
Courtesy: American Bantam Association.

Black Sumatra Bantams.
Courtesy: American Bantam Association.

versus buying pullets, realize that each pullet, by the time it starts to lay, will have consumed fifteen to twenty pounds of feed. If feed costs 9 cents per pound, feed cost alone is $1.35 to $1.80; add the 40 to 60 cents for the chick, and your cost is up to $1.75 to $2.20 per pullet, not counting equipment, labor, or overhead expenses. Obviously, local costs of feed, chicks, and pullets will affect the decision of which is the more economical pathway.

Ready-to-lay hens are nice because they're (guess what?) ready to lay; given the right conditions, you'll start getting eggs right away. You can buy pullets from commercial enterprises or from people with backyard flocks, for $2.00 to $3.50 for an average bird; fancy breeds are more expensive. The big drawback to ready-to-lay pullets is that they're often hard to come by; but if you can find them, you do have instant eggs.

Yearling hens being culled from flocks are cheaper to purchase (approximately $.85 to $1.75), but they will be laying fewer, although often larger, eggs than the younger birds. Don't think that because a commercial egg producer is culling the older birds they have no more production time left; they can continue to lay well for as much as another year. Their slightly lowered rate of lay is uneconomical for commercial producers, however, although it's perfectly adequate for the backyard flock.

Assuming you now have pullets that are laying, regardless of how you got them, you must consider the basics: housing, feeding, culling, and miscellaneous problems. Some basic background in the biology and physiology of the hen will be helpful in understanding its requirements.

226

Anatomy and Physiology of the Hen

Since the chicken isn't a mammal, and most people are familiar with mammalian, rather than avian, physiology, a quick summary seems to be in order.

Starting from the inside and working out, it's interesting to note that the skeleton of the chicken (like all birds) has many bones that are pneumatic—that is, hollow and connected to the respiratory system. In fact, with its trachea closed off, a hen can breathe through a cut portion of the bone in the wing (humerus). Certain bones also serve as a source of calcium for birds in lay.

Since chickens have no sweat glands, regulation of body temperature must be accomplished by other means. Radiation, convection, and conduction are responsible for heat loss, with much of the loss occurring from the head. At high ambient temperatures, evaporation of water from the respiratory tract acts as an important means of heat dissipation (yes, hot chickens pant).

Chicken skin is fairly thin (as all chicken eaters know) and can be various colors, depending on the presence or absence of various pigments in different skin layers. Carotenoid pigments (which come from eating greens) in the epidermis, when no melanic pigment is present, will result in yellow skin and leg shanks. Melanic pigment in the epidermis will produce black shanks. Yellow in the lower dermis layer will be obscured by black in the epidermis layer, but the reverse will produce light green shanks. The absence of any pigment altogether causes white shanks.

Next comes the most obvious thing about a chicken: its feathers. Feathers help provide warmth and protection and are essential for flight; they are renewed annually. Although when looking at a hen you may just see lots of feathers, there are a specific number of large wing and tail feathers. These feathers are lost at molting and then replaced in a specific order; this order of feather replacement is one way to determine the laying stage of a hen.

The hen's digestive system is relatively short and can best handle a diet low in roughage and fiber. The digestive system starts with the beak and mouth, where salivary glands lubricate the food as it passes to the esophagus. The crop is a sac in the esophagus which acts as a storage compartment. At the end of the esophagus is the proventriculus (true stomach), where hydrochloric acid and pepsin are secreted, initiating protein digestion. The gizzard, with one opening from the proventriculus and one into the small intestine, grinds and crushes the food, assisted by grit or gravel eaten by the bird. The gizzard is incredibly strong (and tough to eat) and is comprised of two pairs of thick, powerful muscles. When ground rations are fed, the gizzard is probably not very important for digestion, but it is the gizzard that enables chickens to utilize whole grains.

Silver Laced Wyandotte, female.
Courtesy: National White Wyandotte Club.

Digestion continues with pancreatic juices in the small intestine breaking down proteins, starches, and fats. The liver secretes bile into the duodenum, which is necessary for the proper absorption of fats. The final splitting of proteins into amino acids and reduction of starches to simple sugars take place in the intestine. The lining of the lower small intestine has a huge surface area, allowing for rapid absorption of nutrients; a hen can digest and absorb a full meal in three hours.

Two sacs, ceca, are found at the meeting of the lower small intestine and the rectum; some digestion of very fibrous material may take place here, aided by microorganisms. A short rectum leading to the cloaca is the total of the large intestine. The cloaca, opening at the vent, is an organ shared by the digestive, urinary, and reproductive tracts. Urine is excreted into the cloaca and passes out with the feces; the white stuff seen in chicken droppings is mostly uric acid (waste nitrogen) precipitated from the urine.

Chickens, as you may have guessed, have a very small cerebral cortex—the part of the brain associated with intelligence. However, the optic portions of the brain are highly developed; chickens are visual and apparently can distinguish colors very well. A sense of hearing is also well developed in hens; just look at the amount of verbal communication in a hen house or between a hen and her chicks. Sense of smell, however, does not seem to be important for chickens.

Single-comb Light Brown Leghorn hen.
Credit: USDA photo.

Although hens are not intelligent animals, their social structure, or pecking order, is well known. Pecking order is a well-maintained system: the top bird can peck any other bird, and lower birds can peck only those below them. The lowest bird can be pecked by all the others and has no right to fight back. Adding birds to an existing flock will necessitate a reestablishment of the pecking order, so it must be done with care.

Now we get to reproduction, the major system of concern in the laying hen.

New Hampshire hen.
Credit: USDA photo.

Birds differ from mammals in having only one ovary and oviduct. The ovary of a laying hen contains five to six yellow follicles—the developing egg yolks—plus numerous undeveloped follicles. Egg white secretion and membrane and shell formation take place in the oviduct.

As the young hen matures, FSH (follicle stimulating hormone) from the pituitary stimulates development of the ovary and the follicles. Eventually, the ovary matures to the point where it secretes estrogen, causing development of the oviduct and an increase in blood calcium, proteins, fats, and vitamins needed for egg formation. Body changes also occur under the influence of estrogen: the pubic bones spread and the vent enlarges. Progesterone, another hormone secreted by the ovary, causes the release of LH (leutinizing hormone) from the pituitary, which in turn causes the release of a mature egg yolk from the ovary into the infundibulum of the oviduct. (The infundibulum is a funnel-like structure at the end of the oviduct. Occasionally a yolk misses the infundibulum and gets lost in the abdominal cavity. A hen that does this often is called an internal layer and appears to be in laying condition without ever producing a usable egg.)

Once the yolk has been caught by the infundibulum, it passes to the magnum, where the albumen, or egg white, is deposited around the yolk. Two shell membranes are added in the isthmus section of the oviduct. The thick outer membrane and the thin inner one adhere to one another except at the large end of the egg, where they separate to form the air cell. The air cell is quite small when the egg is laid, but as the egg cools and ages, the cell expands; one way to tell if an egg is old is by the size of the air cell.

In the uterus, the eggshell is formed, composed almost entirely of calcium carbonate, and is bound onto the shell membranes. Eggshell formation necessitates a very high dietary requirement for calcium. This entire process takes about twenty-seven hours, the egg remaining longest in the uterus.

Variation in egg size is common but not well understood. Pullets just starting to lay are thought to lay small eggs because small yolks are being ovulated. Double-yolked eggs are assumed to be the result of two yolks being ovulated simultaneously (twins). The color of eggs (usually brown or white) is determined by the deposition (or not) of pigment in the shell as it is formed in the uterus. Abnormalities in the shell can be caused by two eggs being present in the uterus at the same time, as might occur if an egg is held up from being layed. If both eggs are then layed at the same time (or close), the second egg may be thin-shelled or may just have its membrane and no shell. (It's always startling to reach into a dark nest box and close your hand on one of those membrane eggs.)

Light Brahma hen.
Credit: USDA photo.

It must be stressed that all this egg-laying activity will start, and continue, without the presence of a rooster. Many new chicken people think that in order to have eggs you must have a rooster, but as you can see from the preceding discussion, this is entirely untrue. In fact, roosters can be a detriment to egg production; they often encourage broodiness in the dual-purpose and heavy breeds. However, although roosters do not bring a hen into production, a very important external factor does: light.

Jersey Black Giant cockerel.
Credit: USDA photo.

When young pullets reach a certain stage of sexual maturity, the increasing day length initiates the release of factors that release the LH and FSH, which in turn initiate sexual maturity and the egg-laying process. If laying begins at too early an age (precocious maturity), the eggs are small; if laying is delayed a bit, the eggs are larger—the reason many hatcheries hatch chicks early in the spring rather than in the winter. Winter-hatched hens will mature when day length is increasing (i.e., December-hatched chicks will mature around April), resulting in precocious maturity. Spring-hatched pullets, however, will mature when light is decreasing in the fall, resulting in larger eggs once the hens start laying. However, decreasing day length will lower egg production, so supplemental light to maintain a fourteen to sixteen-hour day must be provided for laying hens as the natural day shortens. Layers should never be subjected to decreasing day length if they are to stay in production. The light provided need not be bright, but it should be there. (Lighting is discussed further later in this chapter.)

Managing the Backyard Laying Flock

Housing

Housing for your flock should be relatively simple and inexpensive. However, for year-round egg production, you will need electricity in the hen area. If you haven't got electricity, or if you want to try out chickens for a short while without a big housing investment, there is the alternative of the seasonal flock; buy older hens early in the spring—hens that have been laying several months to a year—and provide basic shelter for them. Keep them until late fall or until it starts to get cold, and then slaughter or sell them: you've got no freezing water, freezing eggs, freezing chickens, or freezing self—or electricity—to worry about. As long as you keep your initial investment low, this is a good way to test the waters of chicken keeping. For those who intend to plunge right into the occasionally icy waters of year-round chicken maintenance, properly set up it's not very difficult, even with terrible winters.

A dual-purpose breed hen housed on the floor (as opposed to in batteries, as in commercial operations) needs about four square feet of space; you can get away with less if the birds have outside space for ranging, but keep in mind the cooped-up winter conditions. A five-by-eight-foot shed will do nicely for a flock of about ten hens. The shed should be of comfortable working height for the humans who have to use it and tight enough to protect the birds from weather extremes. Temperature extremes must also be considered; extreme heat (over eighty degrees) can be even worse for chickens than extreme cold. Be sure to provide ample ventilation during the warm months. The effects of freezing temperatures can be reduced by insulation, but as insulation can be expensive, this is an economic decision each flock owner must make. We keep our flock in a large coop in one corner of a big, old barn. For the winter, the coop size is reduced, and hay bales are stacked above the coop and along the full length and height of the two inside walls; the two outside walls are not insulated. Our chickens are definitely cold, but they always manage to lay quite well through the winter, so they must not be really suffering. When insulating against winter cold, realize that chickens give off a tremendous amount of moisture and need a constant supply of fresh air. Adequate ventilation is essential, even during the winter.

Many publications about chickens go through scary descriptions of hen house flooring—concrete, raised wood, wire, etc.—making you feel that if you've got a dirt floor, all will be lost to rodents, predators, and disease. There is no doubt that a rodent- and predator-proof floor is desirable, but a rodent-proof floor also means rodent-proof walls at least a foot up—and it all means money if you've got to install it new (another economic decision to be weighed). We've housed our chickens for years on a dirt floor, which we periodically dust with lime and Rotenone (lime for drying and sanitizing; Rotenone as a lice preventative), and have not had any rodent or disease problems. The trick, however, is to make the area as unappealing to rodents and bacteria as possible. Keeping the flooring clean and dry will cut down on bacterial growth; properly installed and filled feeders, good sanitation, and prompt egg collection will discourage rats. Keep a sharp eye out for signs of rats— eaten eggs, holes, rat droppings, or the rat itself—and nip the problem in the bud immediately (see Chapter 11).

Bedding or litter can be any of several materials (shavings, sawdust, straw, etc.) but must be dry, clean, and capable of absorbing moisture. With a deep litter (at least four to six inches) of sawdust, shavings, hay, or straw, you can clean the coop out once or twice a year, as long as the upper layer of litter is kept dry and is replenished now and then. Hens scratching over the litter will help turn it and keep it dry.

If you're starting a building from scratch, keep it as inexpensive as possible (without being cruel to the birds, of course), using exterior grade plywood for walls, floor, and roof. The building can be attached to skids so it can be moved to different locations, or it can be permanently set with a concrete floor. If you're putting in a wooden floor, raise the building on blocks at least 12 inches to minimize dampness and rodent problems.

Provide outdoor space for the hens in the form of a caged outdoor run, which will give them fresh air, exercise, grit, and minerals from the soil; or allow them free range. If the hens are to be allowed to run free, keep them cooped in for a few weeks at first so they learn where home is and where to lay their eggs. Then let them out during the day only. Gathering them in again at night, if they don't come in automatically, is greatly facilitated by teaching them to come when called. This lesson is easily accomplished by calling them every time you feed them, reinforced by tossing them some treats every now and then, calling them at the same time. Our chicken coop has a chicken-sized hole in the wall that allows the chickens to come and go as they please during the day; at night, the hole is closed off, restricting the chickens to the coop.

When hens are kept year round or when natural light is not available to the coop, supplemental light is necessary to provide a fourteen-to-sixteen-hour day in order to keep the hens in production. A twenty-five- to forty-watt bulb for a forty- to two hundred-square-foot area, set up with a timer, will work well. The cost of maintaining the light is minimal compared to that of maintaining a flock of nonproducing hens through the winter.

Equipment

Most of the equipment used for chickens can be easily made, (cooperative extension offices and some feed stores distribute plans for small-flock equipment) or can be purchased used.

Feeders, which can be the trough or hanging type, should provide approximately three inches of feeding space per bird. They should be large enough to supply a day's feed when one-third to one-half full; to avoid feed waste, feeders should not be filled beyond this point. A lip along the side of the feeder will also help prevent waste. If the lip can be adjusted to the same level as the hen's back, this height, too, will help reduce waste. I prefer a hanging feeder, hung at the level of the bird's back and at least a foot away from any vertical surfaces to discourage rodents.

Waterers must supply plenty of clean water at all times, and be easy to clean. Any type of waterer should be placed on a wire mesh water stand to reduce damp litter in the area. An easily made bucket waterer is illustrated in Fig. 9-2.

Both feeders and waterers must be made so that the chickens cannot roost on them, and they must be located away from roosting and perching spots to avoid droppings in the feed or water.

Perches will be much appreciated by the hens. They should be about two by two inches in cross section and long enough to provide eight to twelve inches of roosting space per bird. If you're setting up several rows of perches, allow ten to twelve inches between each, starting at about two feet from the floor. Since the birds will use the perches quite a bit, a fair amount of droppings will accumulate below, which the birds should be kept away from. One method is to place a removable board under the perch to be cleaned weekly. An alternative is to tightly enclose the area under the perch, so the birds can't get in, and clean it out about once a year.

Nests, of course, are where the action is. One nest, fourteen inches wide, fourteen inches high, and twelve inches deep, will serve three or four hens. An orange crate with a board nailed across the bottom of the opening makes a good two-nest set. A small perch across the front of the nest will allow easy access. Nests should be set at least sixteen inches off the floor. Another nesting setup is the community nest, where all the birds lay in one place. One forty-eight-inch-wide by twenty-four-inch-deep box is large enough for about thirty-five layers. With any nestbox system, discourage the

Fig. 9-1. Two Styles of Chicken Feeders. The covered one is designed for range feeding.

birds from perching on top of the box by sloping the roof. Nests
should be supplied with two inches of nesting material (hay, straw,
shavings, etc.), which will have to be replenished periodically.

A *dust box* is certainly not a necessity, but it is a treat for the
hens. Any old box filled with ashes and kept on the floor of the coop
will be well used by the hens; they're such active dusters, though,
that it will be necessary to refill the box often. Actually, any dusting
material can be used in the box (sawdust, for example), but wood

Fig. 9-2. A waterer using a bucket supported on a platform 12
inches high and 24 inches wide, with plaster laths spaced
1½ inches apart.

ashes help to control lice, so you're . . . well, killing two birds with one stone isn't quite appropriate, but you get the point. If, by the way, your hens have free range, they will probably find dirt areas of their own suitable for dusting. If you've never seen a bird dust itself, by the way, it's quite a sight. They lay in the dust and kick and flap, working the stuff into their feathers. The first time I saw it I didn't know what I was seeing and it gave me quite a scare. A bird who had dusted itself was lying in the dirt on her side with her wing all spread out; I thought she was dead. Then she righted herself a bit, and started twitching and kicking; then I thought she was having a fit and was about to die. It wasn't until later that I realized she was indulging herself in a sunning and dusting spree.

Feeding

Feed and water should be available to the hens at all times and must be clean and free from droppings. Water should be changed daily; don't just keep adding water to a dirty, half-full waterer. Most important, fluid water must be maintained throughout the year; little electric heating elements or heated waterers may be necessary during the winter. Chickens will drink a surprising amount of water: one hundred chickens will drink from five to ten gallons of water per day, depending on environmental temperatures.

Water is water—as long as it's clean and fluid, that's about

Fig. 9-3. Two easily homemade nest boxes. The "community nest" shown will serve up to 35 hens. Two separate nests made from a divided orange crate will serve six to eight hens.

Fig. 9-4. Interior of a chicken coop, showing a perch, nest boxes, hanging feeder, and a waterer on a screen platform.

it—but feed is a whole different ballgame. Chickens, because of their essentially simple-stomached digestive systems, and layers in particular, because of their intense production, need carefully formulated and balanced rations. Also, chickens have the added quirk of having their feed intake determined by the energy content of the feed. That is, a hen consumes feed until she has consumed a certain amount of energy, not until she's consumed a certain amount of feed; she'll eat more of a low-energy feed than of a high-energy one. Therefore, since a hen consumes only enough feed each day as will meet her energy requirements, the protein values in the feed must be such that she receives in that daily feed intake the right amounts of each of the essential amino acids, plus the other nutrients necessary for egg production. It's all quite complicated, and poultry scientists have been working for years on rations and requirements. The wise poultry owner will take advantage of the research and buy commercially mixed laying mash.

There are basically two types of feeding systems: the all mash or the mash plus scratch. All mash is just what it says: the birds continually have a quantity of laying mash available. With mash plus scratch, the amount of mash is reduced, and a scratch feed (such as cracked corn) is scattered so the hens will stir up the litter. A 16 to 18 percent laying mash is common, 16 percent being used for the heavier breeds and 18 percent for smaller Leghorn types. Hens tend to eat most heavily in the morning and in the afternoon, so make sure they have feed available at those times; the heavier breeds eat more than the lighter, and all hens will tend to eat more as ambient temperatures drop. Each hen will consume about one-quarter pound of feed per day, or ninety pounds per year. Another gauge of feed quantity, helpful in judging how well your hens are doing, is that hens should consume about four to six pounds of feed per dozen eggs produced.

Now this is the standard feeding regimen, but what if your conditions are different, as for example, with a range setup? Even with range, laying mash should be available. The advantage, however, to range hens is that they will eat less mash if they have access to good range (not just a bare earth exercise yard). Encourage the hens to scratch in the dirt and grass by throwing down corn and cutting back slightly on the amount of mash available. However, don't assume that your hens will perform well on just a cracked corn and range diet, but *do* assume that such a diet will cut down on mash consumption and consequent costs.

The same holds true for table scraps. Scraps can be fed, but in addition to, not instead of, the mash. Not all scraps have value for poultry, anyway, and spoiled foods are particularly unsatisfactory. Give them only as much scrap as they'll clean up quickly (within an hour)—which will also reduce the attraction for rats. If your hens are confined, summertime lawn clippings will be a welcome addition to the daily fare. Dairy by-products, such as whey and buttermilk, have some value for hens, but again, don't feed these in lieu of water or feed; put some in another dish and leave it a day; then remove what is not consumed.

Egg Production and Handling

Egg production levels vary widely with management and breed and age of bird, but it is natural and necessary to want to know how many eggs a hen lays. From the discussion of reproduction, you can see that a hen cannot lay an egg per day. A very high-producing hen, at top form and management, will lay six eggs per week, but the home flock owner would do well to make calculations based on five eggs per week per good laying hen. A dozen birds, therefore, will provide about five dozen eggs per week.

The length of the laying period is also variable, but most hens will lay well for at least a year, and many for eighteen months or more. Commercial egg places generally don't keep layers in production for more than a year, getting rid of their entire flock after twelve to fifteen months. The home flock owner can usually keep birds longer. The hens will go through a molt in the fall, but under the right conditions, hens can still be quite productive in their second season. Beyond two seasons, economics usually dictate chicken soup, but when you've got only a few birds, individual lives can be spared.

Eggs should be gathered twice a day, both for maximum retention of freshness and to keep hens from learning how delicious eggs are. A hen can (and will) learn to eat eggs, and the habit spreads rapidly through the flock. Be sure to remove any broken eggs you might find, for this reason. Eggs should be cooled (forty-five to fifty-five degrees), and then washed in warm water. There is no need to discard cracked eggs; it is primarily the membrane that keeps eggs clean, so as long as it is intact and the shell is clean, the egg is perfectly edible. Eggs can be stored in the refrigerator for quite a while, although for long-term storage, enclosing the egg cartons with a plastic bag is recommended. Eggs should be stored large end up, which is usually the end with the air cell. Contrary to popular belief, blood spots do not indicate a fertile egg, as evidenced by the fact that even without a rooster you'll find some eggs with blood spots. Blood spots are a haphazard occurrence during ovulation, and although in commercial egg places eggs with blood spots are discarded, there is no reason not to use them.

Molting

Once a year, usually in the fall, hens will go through a molt: the replacement of feathers. There are many misconceptions concerning molting, such as that cessation of laying causes molting and that a hen can't molt and lay at the same time. Neither are true. In fact, hens bred for high production often do continue to lay while they molt. Often, too, a high-producing hen will lay as late into the fall as she can, and so will molt late. A low-producing hen may cease production as early as August, however, which is earlier than is really needed for feather replacement.

Hens usually molt at the head first, then the neck, body, wing, and tail. There is great variation in how long it takes a hen to shed her feathers, but it takes at least six weeks to grow new ones.

The fact that the primary wing feathers are shed and replaced in a specific order can be used to determine the stage of the molt. The secondary wing feathers are the ones closest to the body; then there is the short axial feather; then the ten primary wing feathers. The primaries drop before the secondaries and proceed in order, starting with the one closest to the axial.

The time and rate of the molt is influenced largely by feeding and management and by the weight and physical condition of the bird. If you have no faith that your birds will molt and lay at the same time, you may want to push the process a bit by force-molting. Shutting the birds in for a day or two with no feed and little water should stop egg production and bring on a molt. Keep the ration low for another week while feathers are shedding; then feed normally to assist the refeathering process. After force-molting, good feeding and management should bring on a steady return to lay.

Culling

In a small flock, especially one that's kept beyond a year, routine culling is necessary. There is absolutely no advantage to feeding a nonlaying hen (unless it's a pet), so if your egg production isn't quite what it should be, it's time to think about pulling out those unproductive birds. There are several ways to determine whether a hen is laying.

1. The vent on a layer is large, moist, and pliable, as opposed to the puckered, hard condition of a nonlayer. On a yellow-skinned bird, a white or pink vent indicates laying, rather than a yellow vent.

2. Other than the vent, areas to check for pigmentation include the eye ring and the beak. Pigment is withdrawn from these areas as a hen lays, the pigment being diverted to the developing egg yolks. A bleached eye ring (inner edges of the eyelid) indicates a layer. Color disappears from the beak, starting at the base and proceeding to the tip. When laying stops, the pigment returns in the same order it was lost. Therefore, a hen with a beak that is colorless at the tip but yellow at the base is out of production. In reverse, though, a hen with a beak that is yellow at the tip but colorless at the base has been laying for a while.

3. The spacing of pelvic bones is a good indicator of laying condition. In a laying hen, the bones just above the vent (easily felt through the feathers) are wide apart and pliable. In a nonlayer, they are quite close together. One finger-width spread of the bones indicates a nonlayer; three fingers is a good layer (see Fig. 9-5).

In addition to culling for reasons of production, any hen that is injured or sick should be culled. Occasionally a hen will be so low in the scheme of things and be so harassed by the other hens that she'll lead a miserable and unproductive existence—a cull candidate for sure. A broody hen (one that sits on a nest all the time) is another destined for the pot.

Problems and Diseases

Careful sanitation should reduce disease, but you can expect a mortality rate of about 1 percent per month from natural causes. Sanitation means keeping the litter dry and the feed and water fresh and clean. It's not much, but it's amazing how many people neglect this simple preventative.

Any chicks you acquire should be (or should have been) vaccinated against Marek's disease, bronchitis, and Newcastle's disease. Other than these, the only diseases of practical concern are coccidiosis and internal and external parasites. The main thing to remember about coccidiosis, a fatal infection of the intestinal tract, is that the organism needs wet litter to grow; again, keep litter dry. Blood in the droppings is the common sign of coccidiosis. Treatment and control by drugs is possible, but careful sanitation is the best measure.

Sanitation is also important in controlling parasite infestations. If parasite-free chicks are started in clean houses with clean litter, infestation is minimal. Worms are more likely to occur where hens of different ages are mixed, so that the reproductive cycle of the parasite is not broken by house cleaning and repopulation. Chickens heavily infested will be less productive and will appear unthrifty. The most common worm is the ascarid (roundworm), which can be controlled with piperazine in the drinking water.

External parasites are more often a problem than internal ones; lice and mites are the most common. Again, good sanitation and not mixing birds is the best prevention. To look for parasites, catch a few birds and spread the feathers around the vent and neck. Lice and mites can be controlled by dusting the hens and the hen house with an insecticide; recommended are Sevin, Malathion, Rabob, or CoRal. Carefully follow directions on the specific insecticide used. The recommendation for hens housed on the floor and infected

Fig. 9-5. Checking for laying condition by measuring the spread
between the pelvic bones. A wide space between the bones
(3 fingers) indicates a layer in full production; two fingers
(as shown), a hen coming into or going out of production;
one finger, a nonlayer.

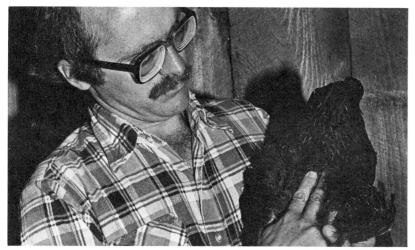

with mites and/or lice is to apply Sevin (5 percent Carbaryl) as a dust on the litter at a rate of one pound per forty square feet of litter, covering as evenly as possible. Repeat in four weeks. Do not put dust into the nests, and avoid contamination of feed or water. Do not keep birds on treated litter within seven days of slaughter.

There are organic methods of controlling parasites, including dusting the birds with wood ashes for lice and painting the lower walls of the coop and the roosts with motor oil.

Predators are another problem for the chicken owner, particularly if the hens are allowed range. Range hens must be taught to come in at dark, and the coop must be made as tight as possible; allow the hens to go out again at 7:00 or 8:00 A.M. It's not only foxes, raccoons, and skunks that you have to look out for, but very possibly the most serious threat will be the neighbor's dog out for a morning of fun. Neighbors must be made to understand that their dogs cannot come onto your property and kill your chickens. Once a dog has killed a hen, it will keep coming back; it becomes a question of your chickens and their right to roam your property versus the neighbor's dog and its right to roam your property. Unfortunately, some neighbors can be made to understand only after their dog has come home full of buckshot.

Fox damage, by the way, can be spotted by the remains. Most predators usually will carry away the kill. A fox, on the other hand, often eats it on the spot and leaves the entrails spread around. If you find an entire bird, dead and mangled, it was, in all likelihood, a dog.

Raising Chicks

Starting Day-Old Chicks

Raising chicks is a lot of fun, and it's a good way to get the kind of hens you want, but it does take somewhat more equipment and care than starting a flock with mature birds. You must realize, of course, that this discussion is on raising purchased day-old chicks. That lovely farm scene in children's books of mama hen with her ten little fluffies is a vanishing one, except in the minds of illustrators. Quite simply, mothering has been bred out of modern-day hens, because mothering instincts and top egg production just don't mix. In order for a hen to become a mother, in addition to being mated, she must be broody. A broody hen is one that just wants to sit on eggs—"set" in chicken world jargon. A setting hen is not laying; she just sets what she's layed already. It's easy to see why commercial egg producers aren't too keen on broodiness. So the geneticists and breeders got together, selected heavily against broodiness, and the trait is gone. However, lest you lose all hope of ever seeing that picturesque scene, the instinct has not been bred out of all breeds. If you want a broody Leghorn, forget it. But some hens of the other breeds can be induced to go broody, notably the Bantams and the heavy breeds: Cochins, Australorps, Rhode Island Reds, and crosses thereof.

Whether raising the chicks for meat, layers, or both, the initial procedure is the same. Be sure to have the whole thing set up and ready to go *before* the birds arrive.

For the first several weeks, a goodly amount of supplemental heat will have to be provided for the chicks. Their comfort range is 80 to 110 degrees, well above the temperature of most of our houses or barns. The heat *must* be supplied; chilling will kill a chick good and fast by paralyzing its breathing apparatus. Its lungs are up along its back, where there's not much protective covering in those early weeks. In a natural situation, the chick, when cold, would go underneath the hen. In an artificial setup, they try to go under one another, pile up, and smother.

A thermostatically controlled brooder—a big metal canopy with a heating element in the middle—is available from farm supply places, but unless you plan to make this an ongoing enterprise or raise a hundred or more chicks at a shot, it's probably too expensive to be economical. An alternative is a heat lamp (or two, if necessary) suspended eighteen to twenty inches above the litter, inside a box two or three feet wide by three or four feet long and fifteen to twenty inches deep; this space will be big enough for twenty-five to thirty chicks for two to three weeks. The heat supply must be such that it will provide 94 degrees the first day (measured at litter level) and can be reduced gradually to 88 degrees by the ninth day, and 80 degrees by the eighteenth day. By six weeks, at the most, extra heat should no longer be necessary. In addition to measuring the temperatures with a thermometer before the chicks arrive, and then checking it periodically throughout the brooding time, you can tell whether there's enough heat by the behavior of the chicks. If they're all clustered right under the light, it's not warm enough; try lowering the light a little or adding another one, if you're using a heat lamp. If all the chicks are spread out at a distance from the lamp, it's too hot. The ideal is to have the exact center space under the lamp almost empty of chicks; then a nice uncrowded ring of chicks right around that, slowly spreading out with a few chicks on the periphery. There should be at least seven square inches of heated brooder space per chick, and as they need less heat, one-half square foot of floor area for the first four to six weeks. Twice this amount of floor area is needed per chick for the second four to six weeks.

Although the chicks need enough space to get away from the heat source when they want to, at first they must be taught where the heat is. Therefore, they should be confined to a smaller area around the brooder for about one week, while still having some unheated space. Usually it can be done by placing a ten-inch-wide band of cardboard in a circle around the brooder, about two feet out from its edge. It is said that at first chicks should be confined in a rounded space; if using a box for a brooder, round off the corners. The reason given is that amazingly enough, a baby chick can get into a corner and not know how to turn around and get out (who said chickens were dumb?). Even more amazing is that the other chicks will also go into that corner, and they'll all pile up and suffocate. I've never seen it happen, but I always use round spaces. Anyway, the guard serves other functions: it keeps the chicks near the heat source until they learn to get to it on their own, and it cuts out floor drafts.

Litter or bedding is also critical. The main criteria for bedding are that it be dry and absorbent and have some insulating value. A three-inch-deep litter is advised. If you're using the cardboard box method, newspaper is okay, but a regular litter is better. Although litter can be almost any material—shavings, chopped hay or straw, rice hulls, dried sugar cane, shredded newspaper, etc.—*don't* use sawdust. Sawdust looks just like food to chicks; they'll eat it, clog up their crops, and die. If you must use sawdust, cover it with newspaper or some other material for the first few weeks. When we raise chicks we use a regular brooder (which we inherited) set up

Fig. 9-6. Chick setup. Chicks are shown in good distribution around the heat (dotted circle). Waterers on platforms and feeders are placed both within and outside of the heated area. Egg flats are used as additional temporary feeders for the first few days.

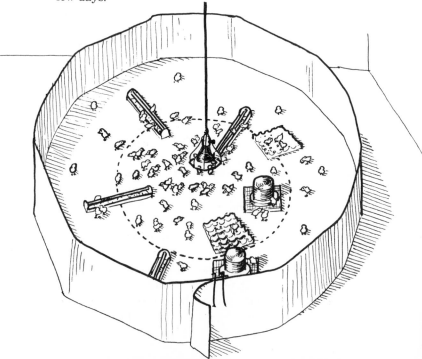

in a room in the barn with a large chick guard around it. We use sawdust for the first two inches of litter; then dried sugar cane for the upper two inches. By the time the chicks are big enough to scratch up the sawdust from below, they are big enough not to eat it.

Chick feeders should be placed so that part of the feeder is within the heated space and part of it out; the chicks can eat in the heat of their choice (see Fig. 9-6). Plans for easily made chick feeders are available from cooperative extension services, but frequently they can be bought used. The essence of the feeder, though, is to provide the feed while avoiding waste. A trough with a lip on the edge and a rail across the top, at a height that allows only heads, not bodies, into the feeder, serves this purpose. Be sure the rail cannot be perched on. Feeders should be filled only two-thirds full, and old feed should be mixed with the new. Use more feeders rather than more feed in each if you want to cut down on the number of feeder fillings. One inch of feeder space per bird is enough for the first couple of weeks; then two inches at three weeks, and three inches at six weeks. Waterers are also easily made; again, avoid investing in new equipment if you can. A one-gallon waterer will suffice for fifty chicks at the start, but obviously more will be necessary as the chicks grow. If a trough-type waterer is used, ten inches will do for fifty chicks initially, increasing to twenty-five inches after a few weeks. Waterers should be placed in several areas, both close to the heat as well as away from it. Placing waterers on raised screen platforms will keep the litter in the surrounding area dry.

All this, including a quantity of chick starter feed, should be set up and tested before the chicks arrive. When you get the chicks, put them right into the brooder. To give them an idea about drinking (and to give them a drink), as you place each one in the brooder, first dip its beak into the waterer. To help them start eating, have several egg flats (or some such) around with feed scattered in them, along with the regular feeders. Because the egg flats are nice and low, the chicks will quickly climb all over them and peck at whatever is there; so make sure it's feed that's there for them to peck. They'll waste most of it, but they'll learn to eat; the flats can be removed after a couple of days. Observe the chicks closely for the first several hours to check their response to the heat source, and adjust it accordingly. Once they're settled in, eating and drinking (and making merry), they should be fine as long as you make sure the temperature is maintained, there's enough feed, plenty of fresh water, and the litter is dry.

A quick word here on feed. As stated previously chicken rations are quite complex, and this is true for chicks as well. Again, depending on the energy content of the feed, protein levels of 20 to 25 percent will be necessary for the first six weeks (starter feed), and 15 to 20 percent up to twelve weeks (grower rations). In addition to being concerned with the energy and protein content of the feed, the physical nature—the degree of grinding—is also important. Actually, because of the increased growth rate of the chicks and the lower feed requirement due to less waste, pelleted rations, if available, are best. The pelleting process apparently improves the digestibility of some feeds as well. Mashes are fine, however, and are certainly more readily available. As you can see, making up your own chick ration would be quite difficult; play it safe and buy commercial chick starter.

As they grow and feather out, the chicks' need for supplemental heat will diminish, and it should not be required at all after six to seven weeks. However, if it's still very early in the spring, hold off throwing the chicks outdoors until the weather is warmer. Raising chicks in the late spring or early summer makes the transition time smoother and faster. Feeders should be adjusted as the chicks grow to keep the height of the feeder at the level of the chicken's back. More feeder and waterer space may be needed, and the litter may have to be turned or renewed more often.

If you're raising layers, low roosts with four to five inches of space per chick can be made available after four weeks. *Don't* put the chicks in with older hens you may already have; wait till the chicks are completely feathered and a few months old. Before mixing chicks into the flock, clean the coop, dust for lice and mites, and administer wormer to the layers. To avoid the older birds excessively picking on the younger ones, be careful how you introduce the two; mixing birds in at night is a good method, as is allowing them to meet outdoors in plenty of space. Housing the young hens and the older ones next to each other (but not together) for a week or so, and then removing the partition one night, is a nice, sneaky way to do it. When mixing the birds together, be sure there's ample feeder space, as the older hens will tend to dominate and try to squeeze the young ones out. One way to distinguish your young hens from the older ones is by color: if you've got black hens, raise red chicks this year and barred the next, so at a glance you can tell just how old a hen is.

If you've raised a group of straight run chicks, the males will have to be separated from the future layers and be slaughtered at around ten weeks. Determining the sex of day-old chicks is possible (by cloacal observation of the rudimentary copulatory organ) but not for the novice. By six or seven weeks, however, sexing by secondary sex characteristics—size, comb, voice—should not be too difficult. Sexing is easy when using sex-linked genes for feather color. A Barred Rock crossed with a Rhode Island Red, for example, produces black hens and barred cockerels. At the day-old stage, they're both black and fluffy, but the cockerels have a light spot on the top of their head.

Rather than buying day-old chicks, it is possible to buy or gather fertile eggs and hatch them, either naturally (using a broody hen) or artificially (using an incubator).

A broody hen can be a bird from your own flock that has gone broody or a Bantam, well known for broodiness and mothering ability. If you plan to hatch eggs from your own flock, separate the rooster and keep him with no more than ten hens. Begin to gather the eggs from these hens two to three days after they've been with the rooster. Collect only clean, unbroken, smooth-shelled eggs of a normal size and shape. Eggs can be collected and stored for a week to ten days before being placed under the hen. Keep them in a cool, dry, frost-free, clean, odor-free, and evenly temperatured (50 to 55 degrees) environment. Marking the date of collecting on the egg will prevent your using eggs that are too old. Turn stored eggs 180 degrees once a day.

The hen or hens chosen to hatch the chicks should have first proven themselves capable of the job. Some hens will set for a few days and then quit; result: loss of the whole batch. You can test the hen by giving her dummy eggs to sit on first; if she resents disturbance and has to be coaxed off the nest to eat and drink, or if she leaves the nest for only a few minutes at a time, then she can probably be trusted with the real things.

The only thing you have to do is set the broody hens up in a quiet, somewhat darkened area away from the other birds and safe from predators. Provide plenty of nesting material (plus boxes, if you like) and fresh feed and water in the brooding pen. Then wait twenty-one days, and the little chickies should hatch out. It's interesting to realize that even though the eggs were laid on different dates, they will all hatch at the same time. The time of egg hatching is actually synchronized by the embryonic chicks themselves; vibrations and clicks made by the embryos will speed up the hatching time of the later-laid eggs. After the chicks are hatched you can separate them from the hen and raise them yourself, which will encourage her to go back to laying sooner, or you can let the hen raise them. Continue to keep the hen and her chicks separate from the rest of the flock for a couple of weeks.

When hatching eggs in an incubator, you are the mother hen, and you are just as tied down as she would be. Temperature is maintained automatically, but you must check it and maintain the humidity, and more important, you must turn the eggs. In a natural situation, the hen turns the eggs frequently; using an incubator, you must turn the eggs six to eight times daily (no going away for weekends while you're sitting on your nest). Eggs are normally incubated large end up and turned only through an angle of 90 degrees. Temperature must be maintained between 98.6 and 100.4

degrees. Embryos can stand accidental chilling best in the early stages of development, but they are highly susceptible to temperature drops during the last three days. Details on the maintenance of temperature and humidity will be supplied with the particular incubator you buy. The advantage to hatching eggs in an incubator is that you are in somewhat more control, and of course, your hens keep laying. The advantage to natural hatching is obvious: let the hen do the work.

Chickens for Meat

Whether you've raised all meat birds or both meat chicks and layers, after six weeks the ration should be changed to a chick grower feed. This feed, lower in protein than the starter, should be used to slaughter size. Slaughter size, of course, is not fixed, but is dependent on economics and what is desired from the bird.

The broiler industry in the United States is truly phenomenal in its efficiency. In somewhat less than eight weeks a broiler will get to a market weight of 3.8 pounds, with a feed conversion of 2.05 pounds of feed used per pound of weight gained. No other livestock industry can match this efficiency of feed conversion. Unfortunately, while speed and efficiency were gained, many people claim that flavor was lost. Some attribute the decline in flavor to the age of the slaughtered birds: too young to have developed tasty meat. Others claim that it's the diet, that when birds were taken off the range, the flavor took off too. Certainly the French, who are much finer connoisseurs of poultry than we are, believe that not only is range essential for a good tasting chicken, but also the *quality* of the range will even further determine the taste of the bird.

I believe that both diet and age influence taste; however, I do place a lot of weight on diet, based entirely on personal experience, not on scientific data. I have eaten eight-week-old range-raised birds and eight-week-old commercial broilers and found a world of difference. To check whether it might all be in my head, I once served a mixture of range and commercial chicken in a casserole to several friends. *I* could tell which pieces were which, but they couldn't, and *everyone* picked the range chicken as vastly superior in flavor. Not proof, by any means, but good enough to convince me that there really is a difference. So, when raising meat birds, I would definitely allow them access to range.

Feeding and Housing

Let the meat birds eat all the grower ration they can consume, and carefully check the setup of the feeders to be sure they are eating with minimal waste. Expect a consumption rate of about seven to ten pounds of feed per cockerel raised to eight weeks. (Feed

Table 9-1. Feed Consumption: Cockerels on a Broiler Test

Age in Weeks	Weight (pounds)	Cumulative Feed Consumption (pounds)
6	3.10	5.40
7	3.80	7.03
8	4.60	9.00
9	5.45	11.34
10	6.20	13.79
11	6.90	16.46
12	7.40	18.98

consumption figures for cockerels on a broiler test are shown in Table 9-1.) If the birds are allowed range, feed consumption may be slightly lower. To provide outside space for the cockerels, a run can be constructed adjacent to the coop which would allow the chicks to be outside during the day and inside at night. Or you can allow the birds free range, but be sure to teach them to come in again at night; birds must be protected from predators from the evening through the early morning. Of course, birds can be allowed range only after their need for supplemental heat has passed. The heat should have been decreased gradually, and birds should have been without any extra heat for several days before going outside.

Housing for the chicks from six weeks to slaughter is the same as outlined previously. Clean, dry conditions and clean feed and water are of paramount importance.

Classes and Grades

The age at which you slaughter the chicken will affect its taste and texture—and what it's called.

Broiler or fryer: nine to twelve weeks old, either sex; tender meated with soft, pliable, smooth-textured skin and flexible breastbone cartilage.

Roaster: twelve to twenty weeks old, either sex; tender meated with soft, pliable, smooth-textured skin and flexible breast-bone cartilage.

Capon: a surgically neutered male less than eight months old; tender meated with soft, pliable, smooth-textured skin.

Stag: a male chicken less than ten months old with coarse skin; somewhat tougher, darkened flesh; and significantly hardened breastbone cartilage.

Hen, stewing chicken, or fowl: a female, over ten months old, with meat less tender than a roaster and a nonflexible breastbone tip.

Rooster or cock: a mature male with toughened, darkened meat, coarse skin, and hardened breastbone tip.

None of these classes is inferior in quality to any other; it all depends on what you want from your meat birds. One of the joys of growing your own is that you can tailor your food to your own taste or to the taste of a specialized market.

Slaughter

Slaughtering is not a pleasant job no matter how it's done, but there are a variety of methods. The fastest, easiest, and cleanest by far is to bring your birds to a place that slaughters poultry. We used to do our own until we found a nearby plant that would slaughter chickens for fifty cents apiece, and have them federally inspected to boot (which meant I could sell them legally). That sounded great to us, so now, unless we have just one or two birds, we bring them there.

Preparing a chicken for the table involves killing, cleaning, plucking, and chilling. Killing can be accomplished either by breaking the chicken's neck or by cutting its head off. (There is also a method of piercing the brain through the upper portion of the mouth, but it is extremely difficult for the novice.) To break the neck, hold both feet in your left hand, with the head downwards; grasp the chicken's head with the right hand, with the thumb behind the skull and fingers underneath the head. Pull downwards as hard and fast as possible, at the same time twisting your hand back so the thumb pushes down and the fingers rotate the head back and up. The head on a broken-necked chicken will dangle visibly, although the bird will continue to flap.

To cut off the head, just lay the bird on something firm, and with a *sharp* cleaver, chop off the head. Again, the headless bird will continue to flap, and would also run, so hang onto it. The carcass must now be bled. Either hang it upside down someplace, or what we do for neatness' sake, hold it over a plastic-lined trash can while it flaps (a minute or so) so the blood runs right into the can rather than all over the ground, where dogs and others would quickly learn what is inside a chicken. (All the rest of the cleanings can then go into the same trash can, which can be covered tightly until removed.)

After the chicken has been killed and bled, I skin it. I know this is a rather radical departure from the tradition of plucking, but plucking takes so long and is such a mess that to invest all that effort just to keep the skin on doesn't seem worth it. I'm not such a great fan of chicken skin, anyway, and no one ever seems to miss it, no matter how the chicken is prepared.

A chicken is skinned much like anything else. Lay the bird on its back, lift up the skin over the abdomen, and cut a slit. Insert your thumbs into the slit, and rip the skin up towards the head, peeling it from the body. Pull the wings through like peeling off a glove (you may have to cut it at the last wing section if you can't pull it all the way through). Pull the skin and feathers (still all in one piece) down the back; pull out the legs as you pulled out the wings (the legs are easier), holding the skin in one hand and the leg in the other, actually peeling off the skin. Again, you may have to cut off the feet when the skin reaches the bottom of the leg joint. You're now left with a wad of feathers and skin attached to the chicken only at the vent. Cut the skin free, which will leave some skin and feathers around the vent to be removed later. This procedure may sound difficult, but after the first time it's easy, and it's much faster and neater than plucking.

If you're bent on plucking, do it immediately after the bird has been killed, since the feathers come out more easily while the carcass is still warm. Keep to a regular order in plucking. Pull out large wing and tail feathers first. Then pluck the breast from the top down to the vent; then the back in the same direction, finishing up with the legs and wings. After plucking, using the blunt blade of a small knife and your thumb, pull out the remaining feather stubs. To remove any remaining hairs, the carcass must be singed over a low flame. A variation of this type of plucking involves scalding the chicken first. Water is brought to scalding temperature (160 degrees) in a large kettle, and the entire bird is dipped in for a moment—which loosens the feathers and makes them easier to pull. An entirely different method, one used more often on ducks and geese, is to dip the entire bird in warm melted paraffin (one part beeswax, one part paraffin), allow it to fully harden on the bird, and then pull off the paraffin; supposedly the feathers come with it.

After being skinned or plucked, the bird is gutted. Open the abdomen carefully, lifting up the muscle and making a slit; avoid puncturing anything but the abdominal wall. Hold the chicken in your left hand and reach up into the body cavity with your right, all the way toward the neck, where you'll feel cord-like structures. Pull, and all the entrails will come out in one piece; extra effort may be needed to get the crop through the opening. At this point, if you were culling nonlayers, you can see if you picked right. The presence of many small egg yolks and eggs in various stages of production will be easily seen in the layer (and will bring on a lot of cursing). Cut a V-shaped section around the vent. Run cold water inside the carcass, working your fingers around the backbone to remove any remaining pieces, especially the lungs, which tend to hang on in there (those are the red structures with the little air sacs). Continue washing the bird until it's all clean. Chill it immediately. If you're planning on freezing it, chill it twenty-four hours first—which allows it to go through rigor, thereby tenderizing the meat.

When trying to figure out how much clean carcass you'll end up with from any given weight bird (if you're trying to calculate for selling purposes, for example), you can figure that 9 percent of the live weight will be lost in killing and dressing; feathers are variable but account for about 5 percent, and blood accounts for the rest. Evisceration losses are also highly variable, but Table 9-2 indicates approximately what you might expect.

Keeping chickens for eggs and/or meat is so easy, so economical, and so much fun, that no small holding should be without a flock of feathered friends.

Table 9-2. Evisceration Losses (in pounds)

AVERAGE LIVEWEIGHT	DRESSED WEIGHT	EVISCERATED WEIGHT	EDIBLE MEAT (WITH GIBLETS)
2.0	1.8	1.4	1.1
3.0	2.7	2.2	1.7
3.5	3.1	2.5	2.0
4.0	3.6	2.9	2.3
4.5	4.0	3.3	2.6

From C. E. Card and M. C. Nesheim. *Poultry Production*. Philadelphia, Pa.: Lea and Febiger, 1966.

References

ADAMS, J. E. *Backyard Poultry Raising*. New York: Doubleday & Co., 1977.

BIDDLE, G. H. *Approved Practices in Poultry Production*. Danville, Ill.: The Interstate Printers and Publishers, 1963.

BUNDY, C. E. *Livestock and Poultry Production*. Englewood Cliffs, N.J.: Prentice-Hall, Inc., 1968.

CARD, C. E., and M. C. NESHEIM. *Poultry Production*. Philadelphia, Pa.: Lea and Febiger, 1966.

ENSMINGER, M. E. *Poultry Science*. Danville, Ill.: The Interstate Printers and Publishers, 1971.

LIMBURG, P. R. *Chickens, Chickens, Chickens*. New York: Thomas Nelson, Inc., Publishers, 1975.

MERCIA, L. S. *Raising Poultry the Modern Way*. Charlotte, Vt.: Garden Way Publishing, 1975.

NORDBY, J. E. *Selecting, Fitting, and Showing Poultry*. Danville, Ill.: The Interstate Printers and Publishers, 1964.

SMITH, P., and C. DANIEL. *The Chicken Book*. Boston: Little, Brown and Co., 1975.

V.
HORSES

10. HORSES

10
HORSES

Whether horses have a place in a book on farm livestock is debatable, but what is not debatable is that more and more people are moving to the country and buying horses. Although some horses are still worked on farms, most are kept for pleasure; either way, they're still livestock, which should be cared for intelligently and humanely. This chapter is geared primarily to the novice horse owner: it will deal only with horse care, not showing, riding, working, equipment, or any of those delightful things that go with owning a horse. I don't care whether your horse is pretty or ugly, whether you ride hideously or well, dressage or bareback—all I care about is that your animal is properly and safely handled, for your sake as well as for his.

Notes for the Novice

You've finally moved someplace where you can keep a horse, so you figure, why not? Riding around the area would be a pleasant form of recreation, and how hard could it be to keep a horse? Or perhaps your child has been pestering you for a pony, and this one just happened to turn up free, so. . . . Then again, maybe you want to farm a few acres without using a tractor, so you've acquired a team of draft horses, even though you don't know much about horses. Perhaps, nicest of all, you've always dreamed of having a horse,

and now you think you can finally swing it. People get involved
with horses for a variety of reasons and become horse owners
through a variety of routes. However, one must be wary of plunging
into horse owning without good reason and without a good deal of
thought.

Horses need a lot of care and attention—at least as much as,
if not more than, other livestock. They need decent feeds (and they
eat a good bit) fed on a steady schedule. They need space to exercise
themselves freely, or they need to be given regular exercise by their
owner. Shelter must be provided, and certain routine health practices must be carried out faithfully. Like any livestock, horses will
occasionally need the attention of a vet; you must be prepared to
call the vet when necessary and you should know that in many
areas, "horse vets" charge more than "cow vets."

Next, realize that horse owning can be an expensive proposition. When dealing with a backyard horse, the cost of the horse
itself is often the least of it, although you will probably have several
hundred dollars tied up in the animal. Feed is expensive, and the
bigger the beast (like draft horses, for example), the bigger the feed
bill. Regular shoeing, worming, and other health care, shelter, pasture, bedding—all thse costs add up, even before the acquisition of
any equipment at all.

These expenses are partly why you should be shy of "free"
horses: the cost of a horse is in its upkeep, not in its initial purchase
price. The horse is free for some reason, after all, and often the
reason is a problem, which will end up costing you more in the long
run in money, time, and heartbreak than you would have spent
buyin; a horse outright at a fair price. If you can't resist a zero
price tag, though, be sure to look your gift horse in the mouth.

Buying a Pleasure Horse

Once you've decided that you want a horse, and you feel prepared
to accept the financial responsibility as well as the labor of ownership, assess your resources. What shelter do you have available?
Will it need modification or will you have to build something new?
Do you have access to feed and bedding? Will you be able to store
the quantities you need, or will you have to purchase feed and
bedding in dribs and drabs (which can be a problem over the winter)? Is pasture available, or will you be buying all your feed? Are
veterinary and farrier services available in the area? Are there
places to ride?

Once these questions and problems have been settled satisfactorily, next think about why you want a horse and what type would
best meet your needs. Learn as much as you can about the different
breeds of horses, their weak as well as their strong points. Are you

interested in a work animal? Obviously a draft horse, draft cross, mule, or donkey is what you'll be looking for. Are you interested primarily in trail riding or in showing or some other form of competition? The purpose for which the horse is wanted should be kept foremost in your mind when buying; there's no point in buying "more horse" than you want or need. Consider the size, disposition and ability of the person or people who will be using the horse; they should be geared to each other. Do not pair an inexperienced rider with an untrained horse—it will be frustrating and unrewarding for all. For the beginner, a gentle, older, well-trained, but not necessarily highly trained, horse is probably best. Small people belong on small horses (usually), and an unaggressive personality should not be expected to handle a fiery beast.

Sources of horses are many; reliable sources of horses are few. Public auctions are the worst possible place for a novice to buy a horse (they're pretty bad even for experienced horse people). Horses are being auctioned because they can't be sold any other way, usually because of some severe vice or problem, which is not at all apparent during the horses' few minutes in the auction ring. There are private and breed auctions, which are safer bets than public ones, but auctions in general are better avoided. Buying from a dealer can also be risky. Ask around about the person's reputation before dealing with him, and bring along an experienced friend to look at any of his horses. People forced to sell their own private horses, because they're moving or going broke or something, are usually a good source; but again, bring along a friend who knows horses, try the horse out yourself a few times, and have it checked by a vet before plunking down the cash. Horse breeders are generally quite reliable but usually have only young stock or quite expensive older stock for sale. However, the extra training that a breeder's horse may have may well be worth the extra cost.

The age and sex of the horse being purchased is another consideration. Stallions are out for inexperienced riders; geldings are usually quieter and more even-tempered than mares; mares have more personality than geldings and can, of course, produce a foal. Horses are in their prime when about seven to nine years old but can be useful for many purposes well into their twenties. As long as the horse and rider's temperament, ability, and training are suitable to each other, age and sex are secondary considerations.

When viewing a prospect, start out by looking at the animal in its stall. Can the owner walk right up to the horse in the stall? Does the horse seem frightened or bad-tempered on being approached (are its ears laid back?)? Has the horse been eating its bedding (is it bedded on shavings but the other horses are bedded on straw?)? Has it been kicking at the walls, chewing on wood, or rubbing its tail? The last, which may show up as a bare spot on the tail or as an abundance of hairs on the wall, may be indicative of a heavy infestation of internal parasites. Has the horse been stall walking or weaving (look for worn-down paths around the edge of the stall or by the door or window)? Any of these vices are difficult to cure, and are at best, annoying, and at worst, dangerous to handler and horse.

Check the horse for limping or lameness as it comes out of the stall. Then watch the horse both at a walk and at a trot from directly behind, directly in front, and from the side. Be sure the horse is moving straightly and evenly. Have the horse circle around you at a walk, trot, and canter, and again check for evenness and smoothness and spring. The horse should move with a lively action, and no foot should hit any other foot (forging or interfering).

Now run your hands over the horse's legs; there should be no strange lumps, bumps, or swellings, and the horse should allow you to pick up each of its feet. Legs should be straight and squarely set; toes should not turn in or out; pasterns should slope at about the angle of the hoof. Remember, a horse is only as good as its feet and legs—no legs, no horse, so pay careful attention in that department.

Look at the overall balance of the animal; do all the pieces seem to fit together smoothly? If you are getting a horse just for recreational trail riding, then a plain looking but sound horse will suit your needs fine. A plain horse, because it often is not a show prospect, may be more easily and cheaply acquired than a big, flashy animal. Some points of conformation, though, can affect the quality of the ride given by any horse. Straight shoulders, too-straight legs, and short, straight pasterns all point to a stiff, uncomfortable horse. Look at the shape of the mouth and jaw (parrot mouth or undershot jaw) and see if the incisors meet; a misshapen mouth can often affect a horse's eating ability.

Fig. 10-1. Parts of a Horse

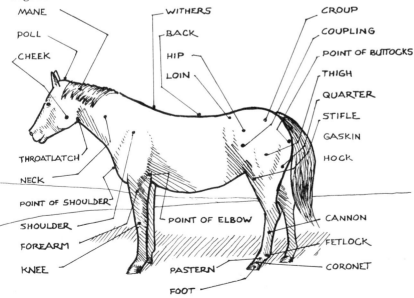

MANE
POLL
CHEEK
WITHERS
BACK
HIP
LOIN
CROUP
COUPLING
POINT OF BUTTOCKS
THIGH
QUARTER
STIFLE
GASKIN
HOCK
THROATLATCH
NECK
POINT OF SHOULDER
SHOULDER
FOREARM
KNEE
POINT OF ELBOW
PASTERN
FOOT
CANNON
FETLOCK
CORONET

Once you've decided you like the horse, try riding it yourself, allowing for the fact that it probably won't behave as well for you as it does for its owner. Ask about papers; you may need a registered animal for the type of use you intend to give it (breeding, showing), or registration may be totally superfluous (pleasure riding) and add only to the price tag. Check vaccination and health background. Finally, see if you can try the horse out another time or two, and if you're satisfied—congratulations! You've joined the ranks of horse owners.

Selecting Work Horses and Mules

Although technically work horses are classified into different categories based on size and weight (draft horses, wagon horses, farm chunks, and southerners), they are all geared to one primary function: work—usually consisting of pulling great weights at a walk.

When looking at a draft animal, keep in mind its primary purpose. Look for size and strength, a deep, broad chest, and a short, strong back. The horse should be well muscled, particularly in the forearm and gaskin, with straight, squarely set, strong-boned legs. Feet should be large with wide heels. As with any horse, consider the temperament and tractability. Watch the action of the horse at a walk and trot; it should be strong, yet energetic.

Mules are the result of a cross between two species: *Equus caballus* (horse) and *Equus asinus* (donkey). A male donkey (jack) crossed with a mare produces a mule; a stallion crossed with a female donkey (jennet) produces a hinny. Either cross is terminal, since both mules and hinnies are sterile.

As strictly work animals, mules have certain advantages over horses. They are easier to manage, being less prone to digestive troubles, more heat tolerant, less susceptible to foot problems, better able to work in low areas, and more careful generally about not hurting themselves.

One would look at a mule much like a horse, checking first for size, strength, and muscling, then for soundness and strength in the feet and legs. Especially when dealing with mules, think about personality; look for liveliness and energy (and grab it if you find it), as well as ease of handling.

Breeds

Over the centuries horses have been selected and bred for certain characteristics, giving rise to the multitude of breeds seen today. For details on specific breeds, write to the breed association and read through breed descriptions in any of the many horse books available. The following is a partial and brief listing of some of the more popular American breeds.

Light Horses (Saddle and Harness Horses)

Thoroughbred: known for speed, stamina, heart, and intelligence.

Quarter horse: originally a cattle working or stock horse, now also widely used for pleasure and on the track (quarter mile).

Morgan: an all-purpose horse with a good disposition; also used for harness racing.

Arabian: gentle and intelligent, known for speed, stamina, and grace.

Appaloosa: usually spotted, with striped hoof; known for endurance and strength.

Palomino: golden-colored horse with light mane and tail; the result of a cross between a chestnut and an albino.

Standardbred: the horse of harness racing.

Tennessee Walking Horse: gentle disposition, gliding gaits (a running walk instead of a trot), making for a very comfortable ride.

American Saddle Horse: comfortable under saddle; also used as a harness horse; often shown as a five-gaited horse.

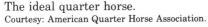

The ideal quarter horse.
Courtesy: American Quarter Horse Association.

An ideal Arabian.
Courtesy: Arabian Horse Registry of America, Inc.

An ideal Palomino quarter horse, by O. Mixer.
Courtesy: Palomino Horse Breeders of America.

Hackney: a superior harness horse, now used primarily for showing.

Shetland pony: a small, rugged pony used as a child's mount.

Welsh pony: a larger pony, used for many riding purposes and considered an excellent pony for more advanced children.

Draft Horses

Clydesdale: known for its ability to pull beer wagons and widely used as a show animal.

Percheron: a popular French, black or gray draft horse.

Belgian: the largest of the draft breeds, weighing up to 2,200 pounds.

Suffolk: chestnut, with superb disposition and a strong desire to please.

Shire: the tallest of the draft horses.

Appaloosa saddle mule.
Courtesy: Bishop Mule Days and the American Donkey and Mule Society.

Thoroughbreds at the Lewis Keller farm in West Virginia.
Credit: USDA-Soil Conservation Service.

Care and Management of the Pleasure Horse

Being around Horses

I'd always heard it applied to snakes and bumblebees, but I was shocked to hear it applied to that huge beast, a horse: "Remember, it's more afraid of you than you are of it. . . ." Yes, placating words, but obviously untrue; how could anything so big be scared of little me? Well, it didn't take long to learn that, in fact, it was true: much of the way a horse acts is based on fear. Those big beasts really do frighten easily—it's built in. Their defenses aren't that great—hitting out with their hooves or running—so they're extremely sensitive to what is going on around them. This is not to say that horses are nervous wrecks, just that sudden, new, or previously conditioned actions can easily cause fear in a horse, seen as bucking, running, snapping, kicking, striking, etc. Most horses do not do these things out of innate meanness but rather from the fear of being hurt, whether that fear is realistic or not. For this reason, always move quietly around your horse, avoiding sudden loud noises or sudden large movements. Always let a horse know of your presence, either by maintaining physical contact with the horse or by talking to it. Be aware that the horse is going to be wary of new things, and that even approaching a familiar object from a new direction can be considered a "new thing." Introduce strange objects slowly, giving the horse a chance to look at and smell them. Many apparently inexplicable bad habits are the products of fear; the

Young donkeys.

horse associates a certain object (your hand, a whip, a jump, a brush) with punishment in its past and continues to react accordingly. So treat your horse gently, realizing why the behavior arose while trying to correct it.

Another trait to be aware of is loyalty; a horse will develop a loyalty to a person and behave differently with that person than with others. The surest way to make sure your horse becomes loyal to you is to feed it. This advice may sound a little crass and not like My Friend Flicka, but it is true that a quick way to a horse's heart is through its stomach. This is not to say that you can simply feed your horse, do nothing else, and expect slavish devotion in return, but just that feeding tends to speed up the bonding process.

Although basically you should deal with your horse in a calm, gentle, and understanding manner, you must also be firm, not letting it develop or get away with bad habits. Tone of voice really communicates a lot to a horse, and a firm, commanding tone used when necessary is an indispensable tool for the horse owner. I bought my mare from someone who had taught the horse to pick up its feet on command, a useful sort of thing. She'd stand by the horse's leg, say "foot!" in a commanding tone, and the horse would pick up its foot. Then I tried it, and it worked. But then I wondered, and went up to the leg and said "avocado!" in an equally commanding tone. Sure enough, up came the foot. The tone of voice, coupled with the stance, not the word itself, conveyed the meaning to the horse. In any event, at the first sign of a developing bad habit, or whenever the horse does something wrong, it must be corrected quickly and firmly and in a way the horse can understand.

Safety Tips

1. Always let a horse know your whereabouts and your intentions. Talk to it.
2. Approach a horse from the left and from the front if possible. When approaching from the rear, let the horse know you're coming; then place your hand on its rump and move slowly up to the head, keeping close to the body.
3. Learn to restrain your horse—cross-tying or holding up one foot, for example.
4. Stand to one side when petting a horse and start by petting on the shoulder or neck. Don't stand directly in front of a horse and try to pet its nose: it can't see you.
5. Walk next to, not ahead of or behind, a horse when leading it.
6. Lead a horse on the left side, with the right hand twelve to twenty-four inches from the bit and the rest of the lead line (or reins) in the left hand. *Do not ever* wrap lead lines or reins around your hand.
7. Don't try to outpull a horse—you'll lose. A sharp snap on the lead line will usually do the trick.

Mention should be made of the horse's range of vision because it also contributes greatly to its behavioral quirks. Horses have what is known as monocular vision: they can see separate images with each eye at the same time. They can also see objects behind them, as long as the object is not narrower than the horse's body. Only when it is interested or excited will a horse strain to see directly in front of it (binocular vision), but even then it is limited to objects at a distance of four feet or more away. In the same way, a horse cannot see directly down in front of itself. Knowing where the horse can see you and where it can't will greatly help in working around the horse and in understanding its actions and reactions.

Shelter

Shelter for your horse can be very simple, but it must provide protection from the elements. It can range from a tight, three-sided shed, oriented away from the wind, to a box stall in a barn.

A three-sided shed in a field is especially good during the spring, summer, and fall. It protects against wind and rain, offers a shady spot to escape from sun and flies, and allows free access. For one horse, the shed should provide at least 120 square feet and be 10 feet high. If they get along well, two horses can be kept in a shed not much bigger than 175 square feet or so, provided they have the freedom to move in and out at will. As winter quarters, a three-sided shed may be less satisfactory. If snow drifts in such a way as to confine the horse to the shed, consideration must be given to space, bedding, exercise, etc., just as would be necessary for a stall in a barn. Also, for winter use the shed must be located in a spot that is accessible and convenient all winter long for feeding and watering. As far as the horse is concerned, if it is a hardy type who will grow a good winter coat, the protection offered by the shed is adequate; so if the other criteria are met, a shed can serve as housing for your horse all year round. There are a couple of other factors to consider, though, when using a shed: you'll need some other place, convenient to the shed, to store hay and feed, and you'll need a place to confine the horse in case of vet checks or illness; tacking up the horse can be cumbersome, especially in bad weather.

A stall in a barn makes excellent living quarters, especially if there is a paddock or pasture the horse can use daily for exercise. A box stall should be at least ten by ten feet (ten by twelve feet or twelve by twelve feet is even better), and the ceiling height should be eight to ten feet. (If a foaling stall is needed, it should be even larger; fourteen by fourteen feet or so, with no hayracks or other protrusions.) The stall should be constructed of at least one-inch thick hardwood, with five-foot-high walls. The boards of the walls must be tight against one another and tight against the floor. A friend of mine once had to have a horse destroyed after it caught a hoof in the space between the bottom board of the stall and the floor.

A well-drained dirt floor, bedded with straw, sawdust, or shavings and regularly cleaned, completes the picture of domestic bliss for the horse. Wooden floors are also used in horse stalls, but the wood can trap moisture and cause odors as well as harbor rodents. Hard-packed clay is extremely popular. Unless it is totally unavoidable, concrete is a poor choice: it is much too hard on the horse's legs and can present serious drainage problems. If you're stuck with concrete, loads of bedding must be used, and it must be replenished often to keep it dry. Covering the concrete with a foot or so of dirt can reduce these problems.

Drainage is a major part of the flooring question; damp conditions are unhealthy and will wreak havoc with a horse's feet, so the flooring and bedding must be such that the floor will remain dry. Bedding, of course, must be kept clean as well as dry; avoid dusty or edible bedding materials. Horse manure mixed with bedding, by the way, is one of the best garden fertilizers, making it quite easy to give away if you don't need it yourself.

Stall doors should open out and be latched low enough to keep the horse from spending its spare time figuring out how to open it. Water and feed buckets located on the front wall of the stall (as opposed to the rear) make feeding and watering infinitely easier. A horse will drink up to twelve gallons of water per day, so make sure the water bucket is of ample size. Rubber buckets are the best bet; ice breaks out of them easily without the bucket breaking. A sixteen- to twenty-quart hard rubber bucket is an excellent waterer, and a second one can serve as a feeder for the grain. Hay can be fed off the floor or from a rack.

If you cannot provide a box stall, a tie stall is okay, but you *must* have an exercise area and you must turn the horse out each day that you don't ride—every day is even better. Tie stalls are usually five feet wide and ten to twelve feet long, with a hay and grain bunk and a water bucket located at the front end.

Any type of stall should be kept dry and draft-free, with no protrusions, loose nails, broken boards, slippery footing, or any other potential hazards. Electricity near the stall is a welcome amenity. Layouts for horse barns are available from extension and plan services (some addresses are listed in the references in this chapter).

Providing the most comfortable quarters as is reasonably possible will greatly pay off in the long run in having a healthier, happier animal. Poor conditions not only run up vet bills but also can affect your horse's disposition, and may lead to vices that are difficult and definitely not fun. So, plan the housing for your horse well and in advance.

Feeding

Compared with what is known about the nutrient requirements of other livestock, we are still in the dark ages with regard to horses. Many excellent researchers have come up with many excellent opinions, but there is nowhere near the wealth of information available on horses that there is on, say, dairy cattle. Also, ask ten different horse people how to feed a horse, and you'll get *at least* ten different answers (a lot of people have more than one theory). However, certain basics have been established; after that, the individual horse owner has to use judgment and the best information available.

General Principles Fresh water must be provided daily, ten to twelve gallons, in a clean container. Heating the water in the winter is not necessary, although it is done by many horse people. An extremely hot horse should not be given very cold water; a hot horse may be allowed a mouthful or two of water, cooled down, and then allowed to drink. Water should be available at all times, but do not let a horse fill up on water just before a stretch of hard work.

Salt must be provided free choice. A salt block, trace mineralized for added insurance, should always be available.

Nutrient requirements of the horse vary with the amount of work the animal is doing. An idle horse (ridden less than one hour per day) needs only a maintenance ration, whereas a horse at hard work (more than five hours of work daily) will need at least twice as much energy. Above all, however, the owner should be aware of the horse's condition in relation to the feeding program. You may be feeding by the book, but if the horse is thin or its coat is dull and dusty, or its eyes are not bright or its attitude is listless, you will have to reassess the ration. Adjustments in the quantity of feed may have to be made if the horse is worked irregularly; rations should be changed gradually. In addition, rations will vary with the productive stage of the horse: growing (foals), reproduction (broodmares and stallions), lactation, etc. However, the individual horse will ultimately decide the feeding level. Some horses just burn feed right up, even if they're merely standing in a stall all day. Others are easy keepers: they can stay sleek and healthy on relatively little feed.

Types of feed for horses fall into two categories: roughages and concentrates. Horses basically eat roughage, that is, hay or pasture. But when hay or pasture is insufficient to meet the nutritional demands of the animal, grain (concentrates) is used for supplementation. Concentrates can range from simple whole oats to commercial sweet feed mixes. It must be realized that as a horse's nutrient needs increase, they are met with increasing concentrates, not roughages. Unlike ruminants, which can handle large quantities of roughage all at once, horses are limited in their capacity at any one time. Therefore, as energy needs increase, as with hard work, lactation, etc., it is the concentrate portion of the ration that must be adjusted.

Quality of the feed is important. Whatever a horse eats must go all the way through its digestive system; a horse cannot spit out or regurgitate feed, nor does it have a rumen to predigest everything. This is why colic is more of a problem with horses than with other livestock; if they eat something bad, they're stuck with it and the problems it might cause. So never, ever, feed moldy feed, be it grain or hay. Avoid dusty, old hay, and don't use dusty or moldy old hay for bedding; most horses will eat it whether you call it "bedding" or "dinner." Sweet feeds should be fresh, as they tend to go rancid if stored too long. Any wet remains of grain left in the feed bucket should be removed daily, since leftovers can mold. Do not feed grass clippings that have been sitting in a bag for a day or two. Grass clippings can be fed to a horse immediately after they've been cut, but not after they've started to ferment.

Feeding schedules must be maintained. Horses like routine and can fret away a fair amount of energy if not fed regularly. Under natural conditions horses graze; that is, they keep eating on and off throughout the day. Providing all day or night grazing is ideal for your horse, but if and when this isn't possible, more, smaller feedings—rather than fewer, larger feedings—is preferable. For the sake of practicality, horses are usually fed their day's ration in two or three feedings; once daily is not recommended. Do not feed grain off the floor, and do not feed grain to a very hot horse. If feeding more than one horse, feed them their grain individually.

Specific Feeding Practices *Pasture* Good, clean pasture is the ideal roughage for horses. A dream pasture is a grass and legume one that is rotated to keep parasites at a minimum, clipped to keep it actively growing, has shade and water available, is free from holes or other hazards, is well drained, and is not too rough or rocky. One to one and a half acres of well-managed pasture—not a weed patch—will supply the roughage your horse needs. If the horse does light work or less, decent pasture may be all it needs, depending of course on the individual type and condition of the horse and how long it is on pasture. When the horse has free access to good pasture, no additional roughage need be provided. If, however, the horse is pastured only part of the day or night, a small quantity of hay (one-half to one pound per hundred pounds of body weight) may be needed while it is in the barn. The hay may be needed not only for nutritional reasons but also to give the horse something to do other than get in trouble. During high heat and fly season, many people pasture the horse only at night, keeping it in the barn during the heat of the day.

Water, shade, and a salt block must be available on the pasture if the horse is to be out all day.

Beware of turning a horse out onto lush green pasture too quickly. Introduce pasture gradually in the spring, and avoid sudden, complete changes to lusher pastures from worn ones or to different types of pasture; any of these practices can throw the horse off its feed.

Fencing for a pasture must meet two requirements: it must keep the horse in and it must be safe. Barbed wire, therefore, is unsuitable. Woven wire is acceptable as long as the holes are smaller than the horse's feet, but it is expensive. Post and rail fencing is very good, very attractive, and very expensive. A well-maintained electric fence, although not the most aesthetic, is probably the most economical and workable type for the pleasure horse owner. Three-and-a-half- to four-foot-high fencing is adequate in most cases. Even though horses can easily jump that height, they're basically easygoing souls who choose the path of least resistance; so as long as there's food in the field, they'll stay in.

Hay Somewhere, the rumor got started that horses can eat low-quality hay, and too many horses are poorly fed as a result. What most horses do not need is the high-quality, all-legume hay fed to dairy cattle; what they do need, however, is high-quality grass or grass–legume hay. Horses simply cannot be fed dusty, moldy, or leached out hay and be expected to do well. Timothy hay is the old standby for horses, as it tends to be quite dust-free, but oat hay, bromegrass, lespedeza, clover, etc. are all good in hay mixes. Straight legume hays tend to be more laxative and more dusty than grass or grass–legume mixes; the best hay for horses is well-cured mixed hay.

As mentioned before, the smaller, slower digestive system of the horse is limited in its capacity for hay at any one time, since in natural conditions, horses graze all day, eating small amounts steadily over a period of time. Normally, horses are fed their hay rations in two or three feedings, the largest quantity in the evening. Anywhere from one to two pounds of hay per hundred pounds of body weight is the rule, but only as a general guide. If the horse begins to develop a hay belly, reduce the roughage and increase the grain. If the horse seems bored or starts to chew wood, add some bulky, but basically nonnutritive, roughage to the ration, such as straw. During cold months, hay digestion offers a good source of heat, so be careful not to feed too little hay during the winter. If you are stuck with dusty (but *not* moldy) hay, you can get away with it by dampening it just before feeding; but it is not wise to depend on this hay for long, since respiratory trouble can develop.

Hay can be fed from a rack, off the floor, or from a hay bag. In my opinion, there is nothing wrong with feeding hay off the floor of a well-maintained stall, but many horse people would disagree violently.

Concentrates The grain portion of your horse's ration can be met with anything from straight whole oats to commercially mixed sweet feeds. Of the cereal grains, corn, oats, sorghum, and barley are the most commonly used. Corn is very high in energy, but low in protein, calcium, and phosphorous, and is considered a heat-producing feed because of its energy content. Corn alone is not a satisfactory feed for horses but rather should be fed in conjunction

Pasture is an ideal roughage for horses.
Credit: USDA-Soil Conservation Service.

with a grass-legume hay and/or a protein supplement. Do not feed ground corn; its lack of bulk will cause colic. Oats have less energy than corn, more bulk, and more protein. As such, oats are frequently fed as the only grain, along with a grass–legume hay. Because of their bulk and ease of digestibility (rolled and crimped are more digestible than whole), oats are unlikely to cause colic. As timothy is the old standby for hay, oats are the old standby for grain. Ground crushed barley is a good feed, but because it tends to cause colic when fed alone, it should be mixed with 15 percent bran or 25 percent oats. Sorghum grain is also fed crushed or ground in mixes with bran or oats.

Commercial mixes are usually a combination of corn, oats, milling by-products, and frequently, molasses. It is often cheaper to buy commercial mixes than straight grain, and they have the advantage of containing the proper vitamin and mineral supplements.

The amount to feed is, of course, the big question. Most horses at light to medium work need from five to ten pounds of grain daily, but this amount varies enormously with the individual horse, the rest of the ration, weather, the type of feed, etc. Table 10-1 is a feeding guide and Table 10-2 shows the daily nutrient needs; use these as guides only. You can see from the tables that normally horses get around two to two and a half pounds of feed per one hundred pounds of body weight, evenly divided between roughages and concentrates.

Do not feed grain off the floor; use a roomy bucket. If the horse bolts its grain, leave a couple of large rocks in the bucket to slow him down.

Complete feeds Complete feeds are pelleted feeds which combine roughages and concentrates in one neat little pellet. The big advantage is obvious: no big, bulky hay bales to handle and store. Also, the ration is completely balanced, no feed is wasted, and the horse doesn't develop a hay belly. The disadvantage is that frequently the pellet-fed horse becomes a bored horse with nothing better to do than think up vices. Feeding straw or coarse grass hay along with the pellets will help keep the horse occupied. One and a half to two pounds of pellets per one hundred pounds of body weight is the usual recommendation for a horse at light work, but (need I say it again?) this guide must be tempered by judgment.

General Health Care

I don't know how to emphasize enough that along with good management, careful observation is vital to both the prevention of disease as well as to the reduction of its severity should it arise. Be aware of your animal. Know its normal behavior pattern, and always see that the pattern is followed. When you go to feed your horse, don't just dump the feed in the bucket and walk away, sat-

Table 10-1. Feeding Guide for Horses

Daily Feed	Grain (lbs. per 100 lbs. body weight)	Roughage (lbs. per 100 lbs. body weight)	Grain Mixes (per 100 lbs. of mix) 1	2
Foals (100–350 lbs.; preweaning)	½–¾	½–¾	oats 80 wheat bran 20	oats 50 wheat bran 40 linseed meal 10
Weanling (350–450 lbs.)	1–1½	1½–2	oats 80 linseed meal 20	oats 70 wheat bran 15 linseed meal 15
Yearling (450–700 lbs.)	1	1½	oats 100	oats 80 wheat bran 20
Two-Year-Old (700–1,000 lbs.)	½–1	1–1½	oats 100	oats 80 wheat bran 20
Mature (900–1,400 lbs.), idle	—	1½–1¾ or pasture	¾ lb. of high protein supplement if hay or pasture is grass only	
light work	2/5–½	1¼–1½		
medium work	¾–1	1–1¼	oats 100	oats 70 corn 30
hard work	1¼–1⅓	1–1¼		
Breeding stallion (in season)	¾–1½	¾–1½	oats 100	corn 35 oats 35 wheat bran 15 wheat 15
Pregnant mare	¾–1½	¾–1½	oats 80 wheat bran 20	oats 95 linseed meal 5

Trace mineralized salt-free choice, mineral mix or calcium-phosphorous supplement free choice with all rations.

isfied that you've done your job. Was there any feed left over from the last feeding? Is the horse coming over to eat with its usual eagerness? Is the horse eating well, both in terms of appetite as well as mechanically (not spilling it from its mouth, for example)? Cast your eye over the horse—is it standing easily? Walking easily? No cuts or scratches? Is it being plagued by flies? Eyes clear? Nose running? All these things take only seconds to observe, but how many problems became serious because the horse owner didn't notice that the horse was slow coming to the barn (was lame) or wasn't eating or was sneezing, etc.? Behavior gives clues to the health of the horse. If the horse is doing something unusual—pawing the ground a lot, for example—ask yourself why, and regard it as a signal. It may be nothing, but it may be the first signs of a disease (such as colic, in this case). Does your horse normally gobble its food right up but is now picking at it? Is your horse usually friendly and frisky but now seems sullen and lifeless? Always be attuned to your horse's behavior, and always ask yourself "why?" It sounds so simple, and it is. It is certainly basic to the caring of *any* animal, but the number of people who don't actually *see* their animals is distressing.

In addition to sensitive observation, basic practices of health management should become ingrained and part of your routine.

Table 10-2. Daily Nutrient Requirements of Light Horses (mature weight, 1,000 to 1,200 lbs.)

	TDN (lbs.)	DIG. PROTEIN (lbs.)	CALCIUM (gms.)	PHOSPHOROUS (gms.)	VIT. A (I.U.)	TOTAL FEED (lbs.)
Weanling (400 lbs., 6 months)	8.5	1.1	33	21	12,000	11.5
Yearling (600-700 lbs.)	9.5	1.7	33	21	16,000	13.5
Two-year-old (800-1,000 lbs.)	10.8	1.4	17	17	25,000	15.5
Mature, idle (less than 1 hr. work daily)	7.1	0.7	12	12	8,000	16.5
Mature, light work (1-3 hrs. daily)	9.5	0.9	24	21	18,000	16.5
Mature, medium work (3-5 hrs. daily)	12.5	1.0	24	21	18,000	19.5
Mature, hard work (over 5 hrs. daily)	15.5	1.3	24	21	18,000	22.5
Breeding stallion (during season)	14.0	1.7	60	40	32,000	22.5
Bred mare, light work	11.0	1.2	24	24	24,000	19.0
Lactating mare	19.0	2.0	40	40	40,000	29.0

1. Always change feeds slowly, including starting the horse on pasture.
2. Feed balanced rations.
3. Feed on a regular schedule.
4. Provide your horse with exercise; work up slowly after periods of rest; wind down slowly after prolonged periods of hard work.
5. If your horse is hot (from work, not from ambient temperature), cool it down slowly. Walk it after you've dismounted, leaving the saddle in place after loosening the girth. Then remove the saddle, and continue to walk the horse. Do not allow it to drink a quantity of cold water; do not give it grain. To see if the horse is cooled, feel behind the forearm or on the chest. A horse can still be wet with sweat but cooled down. Keep it out of drafts while it is wet.
6. Avoid potential dangers. Be aware of loose boards, holes, loose wires, protruding nails, glass, etc. Think ahead about possible problems: store grain in areas inaccessible to the horse; have first-aid equipment handy; know the phone number of a vet.
7. Vaccinate regularly for tetanus (annually) along with any other vaccinations suggested by your vet. Note: horse owners themselves must keep up on their own tetanus shots.
8. Worm twice yearly, or follow a program of internal parasite control suggested by your vet.
9. Groom your horse regularly; be aware of signs of external parasites.
10. Pay attention to, and take care of, your horse's feet—no feet, no horse.
11. Know first aid.
12. *Promptly* attend to all problems.
13. Provide sanitary and healthful living quarters.

None of these practices is difficult, expensive, or time-consuming, yet the difference between doing them and ignoring them can be the difference between having a horse who is a pleasurable, useful animal and one who is a bother and a bore.

Grooming and Foot Care

Grooming is an important part of health care. In addition to keeping the horse's coat clean, stimulating the skin, and reducing external parasites, regular grooming provides an excellent opportunity for the horse owner to really look at the horse closely. It is during grooming that you'll notice small nicks and cuts that should be attended to; bot fly eggs, which should be removed; coat condition and general condition of the mucous membranes; weight; any sore spots, hot spots, swellings, or chafed areas. Daily grooming is the ideal, but it is not absolutely necessary; three or four weekly groomings will do for an idle horse. When working the horse, it should be groomed before being tacked up, especially in the area of the saddle and girth. Dirt in these areas, if not brushed out, can cause chafing and saddle or girth sores. After a ride, a thorough grooming is called for, starting with brisk currying over the entire neck, body, and upper legs, followed by overall brushing to remove the loosened hair, and then rubbing with a soft brush or cloth. The mane should be combed out and the tail brushed, and throughout a sharp eye should be kept for any cuts or sore spots that may have developed during the workout.

After each ride, the feet should also be taken care of; if the horse isn't being ridden, the hooves should still be cleaned several times a week. The foot is picked up and the hoof cleaned out with a hoof pick in order to dislodge any stones and remove the caked-in material. Learn how to use this tool from a knowledgeable friend, a vet, a blacksmith, or a horse-care book. The prime foot problem for the horse owner to be aware of is thrush. Thrush affects the horse who is kept in damp conditions, such as a damp stall; it rarely occurs in horses on pasture, unless the horse is forced to stand in mud for prolonged periods. It makes itself known by smell; when you clean out the horse's feet, you will notice a particularly bad odor if the horse has, or is getting, thrush. At the first sign, rinse the foot with either a commercial thrush rinse or with bleach. Next, be sure that the horse has a clean, dry area to stand in. Continue cleaning out the foot gently every day until it seems to have cleared up. In more advanced cases, the frog of the foot will be black and peeling and may have a discharge. You can't miss the odor on this one—it's vile. At this point, treatment will involve cutting away the diseased tissue and probing for deeper infection; it is a job for your blacksmith.

Foot care does not end with periodic hoof cleaning, however. Depending on the individual horse and the type of work it is doing, as well as the working and living conditions, the horse will have to have its hooves trimmed and shod every four to eight weeks. Some horses, with good feet, ridden lightly on soft ground, may not need shoes at all; the same horse ridden hard over rough, stony ground may need shoes every six weeks. The average horse ridden under average conditions (if there is such a thing) needs shoes every six weeks. Frequently, shoes are removed if the horse is idle, especially during the winter. But the hoof is always growing (although it does grow more slowly during the winter), so it will need regular trimming, even without shoeing.

As soon as you get your horse, ask around about the local farriers and get one to come out to look at the condition of the horse's feet and make recommendations. Hoof trimming and shoeing is one of those things that many backyard pleasure horse owners tend to ignore. Farriery work is expensive, and there's an attitude of, "Well, horses in the wild don't wear shoes. . . ." True enough, but horses in the wild also aren't ridden, aren't kept on hard ground which breaks up their feet, or in soft stalls, which doesn't keep them properly worn down. Also, wild horses that went lame because of poor feet were naturally negatively selected, unlike now, when weak-footed horses are kept, ridden, and bred. A horse with overgrown hooves is uncomfortable and sore and simply cannot move properly. In the extreme, overgrown hooves are a pitiful thing to see—the animal can't even stand adequately, much less move right, its poor feet and legs are so out of whack. Yes, a farrier's work *is* expensive, but owning a horse is expensive, and if you can't afford to take care of your animal properly—and that includes foot care—then you shouldn't own it at all.

Diseases

In all cases, call your veterinarian.

Swamp fever (equine infectious anemia, E.I.A.)—virus

> Symptoms: sporadic high fever, stiffness in the rear, anemia, jaundice, edema, swelling of lower legs and body, loss of condition, death within two to four weeks
> Treatment: none
> Prevention: dispose of infected animals; take sanitary measures; no preventative vaccine

Equine influenza (shipping fever)—myxovirus

> Symptoms (more common in young animals, symptoms develop two to ten days after exposure): rapidly rising high temperature; high temperature for two to ten days; loss of appetite; weakness; listlessness; rapid breathing; cough; watery eye and nasal discharge, changing to a white or yellow discharge
> Treatment: keep animal quiet; antibiotics
> Prevention: sanitation, vaccination, isolation of all new animals for three weeks

Distemper (strangles)—bacteria *Streptococcus equi*

> Symptoms: listlessness; loss of appetite; high fever; nasal discharge; high respiratory rate; swelling of lymph nodes under the jaw, followed by abscess
> Treatment: careful nursing; rest in dry, draft-free stall; fresh drinking water; good feed to encourage eating; possibly antibiotics or sulfa drugs
> Prevention: avoid contact with or contamination from infected animals

Encephalomyelitis (sleeping sickness)—virus, carried by birds, mosquitoes

> Symptoms: lack of coordination, sleepiness, poor vision, grinding of teeth, inability to swallow, fever, paralysis, death within two to four days

Treatment: careful nursing; possibly some serum treatment

Prevention: vaccinate all animals in areas where the disease occurs

Tetanus (lockjaw)—bacteria *Clostridium tetani*

Symptoms: associated with a wound; incubation period of one week to several months (high mortality); stiffness about the head, third eyelid draws over eye, spasms when animal is disturbed

Treatment: good nursing, antitoxin, tranquilizers, antibiotics

Prevention: annual vaccination of all horses is recommended, as tetanus is extremely widespread

Azoturia (Monday morning sickness)—metabolic

Symptoms (occur in animals normally worked hard, rested a day and fed well, and then put back to work); lameness occurs upon resumption of work, with stiffness, sweating, and swelling and tightness of affected muscles

Treatment: immediate rest in standing position, blanket, sedatives, laxatives; do not move animal any distance

Prevention: on idle days, decrease grain feeding by 50 percent and allow exercise; warm up slowly after periods of rest

Laminitis (founder)—metabolic

Symptoms: painful lameness in affected feet, hooves warm, fever, sweating; in chronic cases, hooves become distorted, with walls showing rings parallel to the hair line; symptoms occur following excessive grain consumption or sudden overconsumption of lush pasture, fast work on hard roads, sudden feed changes, drinking cold water while hot, metritis

Treatment: cold packs on feet or standing in cold water or mud puddles; keep feet trimmed

Prevention: good management, particularly with feeding

Colic—digestive

Symptoms: restlessness, obvious discomfort, biting at flank or sides, rolling, sweating; caused by moldy feeds or hay, drinking cold water when hot (from work), riding after full feeding, inadequate water supply, irregularity in feeding, overfeeding, heavy grain feeding directly after work

Treatment: remove feed and water, keep animal quiet, prevent from rolling (walk the horse if necessary), drench with one pint linseed oil plus eight ounces of water or with other colic remedy

Prevention: good management, avoidance of spoiled feeds

Foot and Leg Problems

The number of problems concerning the feet and legs of horses seems endless; it almost seems miraculous that there are any sound horses at all. Problems include spavins, splints, sidebones, navicular disease, ringbone, wind-puffs, stringhalt, and more. Some are serious, and the prognosis is poor; others are mild and require only a small change in exercise patterns or shoeing; but most first show themselves as lameness. Lameness should always be attended to, whether by you, a vet, or a blacksmith. See if you can locate the site of the problem—muscles, hoof, front, rear, etc. Do not ride a lame horse. If the problem does not seem severe, keep the horse under close observation for a few days; the condition probably will get either better or worse. If it clears up, well and good (although you should try to figure out what may have caused the problem in the first place and correct it if possible); if it worsens, call the vet, or if it seems to be a hoof problem, the blacksmith.

Parasites

Parasites affecting horses include the external (flies, lice, mites, ticks) and the internal (strongyles, ascarids, stomach worms, pinworms, and bots).

External parasites are usually just nuisances rather than dangerous, but don't be fooled into thinking that nuisances cannot *become* dangerous. A horse continually plagued by external parasites, even if they're only house flies, can quickly lose condition and become unthrifty and miserable looking (and feeling). The biting flies—horn flies, stable flies, deer flies, and horse flies—in addition to being extremely irritating, can carry serious diseases. Flies associated with wounds can also be dangerous, slowing or actually preventing healing. Keeping wounds clean will reduce this problem, and keeping shelters dry and clean, particularly during the warm seasons, will control flies in general. Pesticides can be used in stables (carefully, and according to the manufacturer's directions), and fly repellents can be applied to the horse daily or as needed.

Lice can be prevented primarily by good grooming, feeding, and general sanitation. They can be transmitted from horse to horse, however, so beware of whom you borrow blankets from. Insecticides are used to treat a horse for lice; consult your vet for the best method.

Parasites vary with locality, so any unusual itching, scratching, unthriftiness, hair loss, or scab formation, which would indicate the presence of external parasites, warrants a phone call to the vet for diagnosis and treatment. Internal parasites require somewhat more sophisticated methods of control. Good management is an essential part of any control program, but even with excellent management, twice yearly worming is recommended. Worms can cause a great deal of damage: poor feed efficiency, poor work efficiency, coughs, bronchitis, colic, anemia, lameness, etc., but the damage occurs slowly and not very obviously. Even if your horse doesn't look like it has worms, routine control measures should still be used.

Some of the following management practices are used to lessen worm infestations:

1. Do not feed off the floor.
2. Provide sanitary drinking areas; drain standing pools.
3. Keep bedding clean.
4. Spread manure in places that will be inaccessible to horses for at least one year.
5. Rotate pastures.
6. Keep stored grain covered to reduce contamination from flies, rodents, and birds.

Generally, horses are wormed after the first killing frost in the fall and then again in the spring. Consult your vet for local practices and specific recommendations.

Foaling

Breeding a Mare

You've got a mare and you've decided to breed her. Choosing the stallion will depend on many criteria, but the first question to answer is why you are breeding the mare in the first place. Is it to get a salable foal, to get a foal for yourself, or just for the experience? Although every attempt should be made to breed to as good a stallion as you can, it is less important if you're doing it only for the experience. If you're aiming to sell the foal, however, in order to make the enterprise at all worthwhile, you must find the best stud you can afford, and at the same time, avoid deluding yourself about how much (if any) money you will make. You also must breed to a stallion who will produce the type of foal most easily sold. If you're breeding the mare to produce a foal for your own use, you should still aim for the best stallion you can find; but instead of looking for economically important traits, as you would if planning to sell the foal, you can look for specific desired characteristics to suit your own needs, possibly even experimenting with crossbreeding. However, it cannot be stressed enough that in general, you should breed up, that is, breed to a stallion who is better than your mare. Always check the pedigree of a purebred stallion.

When picking a mate for your mare, you should be concerned not only with the quality of the stallion but also with the quality of the breeding establishment. Check out the quarters where your mare will be staying and be sure they meet your standards (which should be high) in terms of safety and sanitation. Actually, the entire establishment should be looked at in the same light; you don't want your mare coming home hurt or diseased. Find out how the mares are bred: by artificial insemination, hand mating, or field mating. A.I. is used only to a limited extent and only at the bigger breeding establishments. It has the advantage of being the safest breeding method for the mare and stallion both, and the conception rate, when the insemination is properly done, is as good or better than natural service. The Jockey Club, however, does not allow the use of artificial insemination with thoroughbreds. Field mating is just the opposite of artificial; mare and stallion are turned out in a field and nature takes its course. Unfortunately, nature can also take its toll; this is the least safe breeding method, and I would never allow it for a mare of mine. Hand mating is the most common method, and it is fairly safe for all concerned. The mare's tail is bandaged and her legs may or may not be hobbled; hind shoes are removed. The stallion is brought to the mare in the breeding area and mates her under somewhat controlled and restrained circumstances. The mare is bred once a day or every other day through her heat, beginning with the third day of the heat period. For your mare's sake (and it may be required by some breeding establishments) the mare should be thoroughly examined by a vet several months before breeding.

Horses are naturally somewhat seasonal breeders, coming into heat in the spring and summer, a time when they're in a gaining condition. However, for racing and showing, a horse's age is measured from January 1, so breeders aim to have foals born as close to the first of January as possible, pushing the breeding season up into the winter and very early spring (February and March). This unnaturally early breeding season is the prime cause of the low conception rate (approximately 50 percent) found in mares. However, if the January 1 date does not concern you, spring foaling (therefore spring or summer breeding) is considered best. Heat periods occur approximately every twenty-one days and last from four to six days. Signs of heat include frequent urination, relaxation of the genitals, a mucous discharge, and "winking," an opening and closing of the vulva. Knowing when the mare is due to come into heat, based on watching for heat periods for several previous months, will help determine when to send the mare to the stallion. You could send the mare to the breeding establishment for a month or more, having the mare teased to determine when she's in heat, but that can be expensive. If you know when the mare is due and send her a day or so before that, board bills will be greatly reduced.

The success of the breeding can be determined by (1) the fact that the mare no longer comes into heat, which is not entirely reliable; (2) rectal palpation 40 to 60 days after breeding; (3) any of a few biochemical pregnancy tests. Gestation is a little more than 11 months—336 days—with quite a bit of leeway on either side. The due date may be calculated by adding two days to the breeding date and subtracting one month.

Before committing yourself (or your mare) to a breeding establishment, look at the breeding contract—and there should be a written contract if you're dealing with a stallion of any quality. The contract should clearly define the terms, including the arrangements if the mare doesn't settle, aborts, or has a stillborn foal. Most contracts guarantee a live foal; that is, the mare is rebred at no charge if she doesn't settle, aborts, or has a stillborn. Many contracts go into further detail on the foal, guaranteeing rebreeding if the foal doesn't live past a certain number of days. With the colored breeds, color may be guaranteed (as with Appaloosas, for example: "guaranteed live colored foal" may be the terms of the contract). However detailed the contract, be sure you have the entire agreement in writing, with the coverage offered commensurate with the quality and fee of the stallion.

Care of the Pregnant Mare; Foaling

The main thing about caring for your pregnant mare is to be sure she has a good diet and gets regular exercise. Avoid letting the mare get fat, but give her the best-quality roughage you can, plus a vitamin and mineral supplement. (See Table 10-1 for guideline rations.) Light work is fine for the mare throughout gestation, but keep the amount and type consistent. On idle days or if the mare is not ridden regularly, provide a large pasture, with shade and water, for her to roam freely. Be aware of the dispositions of other horses who may also be in the pasture, and segregate the mare if necessary—this is not the best time for fighting or kicking. However, do not confine the mare to a stall or small paddock during her pregnancy; she must be free to move around.

Signs of approaching parturition include:

1. distended udder, two to six weeks before foaling
2. drop of belly, seven to ten days before
3. loosening of muscles and ligaments around the tailhead, seven to ten days before
4. teats filling, four to six days before
5. loosening of the vulva
6. restlessness, frequent urination

Any, all, or none of these signs may show up, so it is generally considered a wise policy to be ready for foaling a month before the due date. Foaling quarters should be established and prepared. If the weather is reliably good, the mare is best left to foal outside in a small, clean paddock, away from other animals. (If she is left in too big an area, you may not be able to find her to help her when and if assistance is needed.) If the weather is poor or no clean pasture is available, a foaling stall, or just a section of the barn, may be used. A foaling stall must be at least fourteen feet square, and larger if possible. It should be well cleaned and empty of mangers, racks, feeders, etc., which could injure the mare or the foal. If the horse is not accustomed to the area, she should be moved into the stall, nights only, a week or ten days before her expected foaling to allow her time to settle down and relax in the new surroundings. Clean the stall daily, providing plenty of good, clean, dry bedding.

Start decreasing the mare's grain allowance a few days before foaling and begin feeding lighter, more laxative feeds, such as wheat bran.

Extreme restlessness, nervousness, turning and biting at the flanks, lying down and getting up, tail switching, frequent urination, and sweating all indicate that parturition is imminent. Many mares do not like having people around while they're foaling and will actually delay foaling until they're alone. (How many times have you heard about someone staying up all night long with their mare, going out for fifteen minutes for a cup of coffee, and coming back to find the foal?) Making yourself scarce—close by, but not visible—will facilitate matters, but you do want to be around in case of trouble.

If delivery is normal—and it usually is—the waters break and the foal is delivered front feet (hooves pointed down) and nose first, the mare lying on her side. Delivery should be complete in fifteen minutes to a half hour. If the mare seems to be taking too long and laboring unduly hard or if the presentation seems abnormal, call the vet immediately.

Once the foal is born, make sure its nose and mouth are cleared of membranes and that it is breathing well; rub it vigorously with a towel to stimulate it if necessary. Dip the navel in iodine; then leave the mare and foal alone for a bit. Once the foal is up and nursing, which should be within two hours, clean the stall and provide some lukewarm water for the mare. From one to six hours after foaling, the placenta should be expelled. If it is not, call a vet. It is important to examine the placenta. Spread it out—it's T-shaped, two horns and a body—and make sure there are no pieces missing. There will be the one hole where the foal broke through, but that should be all. Pieces of placenta, as well as the whole placenta itself, if retained in the uterus will cause infection and possibly laminitis. If you don't feel up to examining the placenta yourself, save it for the vet to see if you plan on having one come to check the mare and foal.

Start feeding the mare lightly after foaling with laxative mashes and feeds (such as bran and oats) for the first couple of days, gradually getting her on full feed within a week or ten days. (Put the grain and hay low enough for the foal to reach, too.) Be careful to look, immediately after foaling, for the bowel movement of the foal, which should occur within four to twelve hours. If there is nothing, and the foal seems listless or is not nursing, either call a vet, or if you feel able, give the foal an enema. One or two quarts of water with a little glycerine or soap at body temperature, administered by a tube, should be adequate. Consult a vet if the enema does not produce the desired effect. Scouring is the opposite of constipation; should the foal start to scour, reduce the mare's feed and consult a vet.

A mare and her foal.
Credit: USDA photo.

Once everybody seems to be functioning normally, you can turn the mare and foal out together (beware of other horses, though) and sit back and enjoy one of the most beautiful sights in the world: a healthy mare with her eager, frisky newborn foal exploring the world.

References

Breeding and Raising Horses, Agricultural Information Bulletin 394, Washington, D.C.: U.S. Government Printing Office, 1972.

ENSMINGER, M. E. *Horses and Horsemanship,* 4th ed. Danville, Ill.: The Interstate Printers and Publishers, 1969.

HAYES, M. H. *Stable Management and Exercise.* New York: Arco Publishing Co., 1968.

Horse Handbook. Midwest Plan Service. Ames: Iowa State University, 1971.

Horsemanship and Horse Care. Agricultural Information Bulletin 353, Washington, D.C.: U.S. Government Printing Office, 1972.

POSEY, J. K. *The Horsekeepers Handbook.* New York: Winchester Press, 1974.

STONERIDGE, M. A. *A Horse of Your Own.* New York: Doubleday and Co., 1968.

ULMER, D. E., and E. M. JERGENSON. *Approved Practices in Raising and Handling Horses.* Danville, Ill.: The Interstate Printers and Publishers, 1974.

VI.
OTHER
MANAGEMENT

11. THE BARN
12. FIRST AID

11
THE BARN

Setting up your barn properly is one of the first steps in running an economical, efficient, healthy, and happy operation. Too many times you see situations in which, because of a lack of planning, people are making things harder for themselves and/or their stock.

In planning the barn, whether dealing with an old one or building a new one, several factors must be considered: overall economy of design, feeding and watering systems, and manure handling.

Economy of Design

Economy of design is about as broad a phrase as you'll ever want to hear, but it *is* meant to encompass quite a bit. *Economy of space* is one aspect; you want to have as little wasted space as possible. Wasted empty spaces fill with junk, and poor planning of pen areas, storage places, etc. leads to inefficiency. *Labor economy* is also of prime importance: the barn should be arranged to function smoothly. Figure out your routines and plan the layout accordingly, *before* hammering a single nail. Where do you enter; where is the feed located; where is it going; what tools should you keep handy for everyday use; where is the manure heap in relation to the pens; where do you need doors, ramps, etc.; where are fire exits? Go through all the steps of your daily chores, mentally rearranging things to save yourself time and energy. Realize that jobs like wheelbarrowing manure are not the most enjoyable part of keeping an animal, so make the distances and the mechanics of it as easy as possible. Consider labor-saving devices in feeding; for example, if you have a long row of pens where all the animals are getting one or two types of feed, and the feed is located at one end of the barn, putting two trash cans full of feed on a wheeled cart or directly in a wheelbarrow and wheeling it down the row will save the many steps of running back and forth for feed. A lack of efficiency in labor wipes out many small farmers. Your time is worth something, and if you plan on running an economic operation, it *must* be an efficient one.

Heating and ventilating can also come under the canopy of overall design. In a cold climate, winter temperatures are certainly to be reckoned with. Many animals can tolerate the cold if they are dry and out of drafts; others are more sensitive. Grouping animals together during the winter will help keep them warm. If your barn has a high, uninsulated roof, consider constructing a loft above the pens, which can hold hay during the winter and at the same time provide some degree of insulation. It may not be the most aesthetic approach, but plastic over the windows or around certain areas will also help keep in the heat produced by the animals themselves. Very young animals often need warmth. Heat lamps in the pen may be necessary (if the pen can be insulated, so much the better), but for heat lamps you do, after all, need electricity. Consider the electrical supply in areas where supplemental heat may be required.

Hand in hand with heating is ventilating; don't let overconcern for the animals' warmth during the winter smother out fresh air. Confined, overwarm, underventilated quarters are damp, smelly, and unhealthy. A clean, dry draftless but cold barn is definitely preferred over a warm, stuffy one. Fresh air evaporates moisture (and there's a lot of moisture given off by most livestock), cuts down on odors, and is generally healthier for all concerned. During hot weather, cross-ventilation is definitely a plus—overheated animals perform poorly, and a hot box of a barn is certainly less comfortable for the humans involved.

Design for *safety*—safety of humans and animals alike. Door widths should be adequate for the size of the animal, as should ceiling heights. Lighting should be sufficient for animals and humans. Pens should be constructed with attention to details; notice spots where feet, heads, or horns might get caught, for example, and think ahead about materials that might be poisonous (leaded paints) or eventually harmful (chewable, splintery softwoods). Passageways should be kept clear of storage items or debris; don't make your cow walk an obstacle course when coming in for milking. Be aware of potential hazards: loose boards, protruding nails, wet, slippery surfaces, pieces of wire, faulty wiring, etc., and take care to fix such things immediately. Last, but very far from least, is fire safety. Plan fire exits and put in water and sand supplies. Horrible as it is, you must think about just what you would do in case of a fire and set up the barn accordingly. I hate to be so hackneyed about all this, but an ounce of prevention really is worth a pound of cure. Maybe even ten pounds.

One last aspect of overall design and its economy is planning ahead for possible future needs. You may be starting off just with chickens but soon plan to add a few pigs, a steer, and a cow; plan your barn space accordingly.

It's much easier to plan ahead than it is to rip out walls you put in just a few months back. For example, remember that keeping dairy animals involves offspring: plan on calving or kidding quarters plus housing for the little ones. Keeping chickens all year round means light: plan on electricity and possibly heat sources in the chicken area. Working out your immediate needs with allowances for future expansion will save time, money, and headache in the long run.

Handling Feed

Feed, whether homegrown or purchased, is one of the most costly aspects of stockkeeping, and feeding is a major labor consumer. Feed must be stored to maintain its quality, yet be easily accessible for efficient feeding. Space needed for feed storage must be taken into account when planning your barn or deciding the number of animals you can house. This computation is particularly important when it comes to hay storage; hay takes up a lot of space, and it's not something you can easily purchase in small amounts. Especially in cold climates, several months worth of hay is usually kept on hand, which can take up a *lot* of space. Table 11-1 shows approximate space requirements for a few common feeds.

Feed must be stored in a manner that will retain its quality. This requirement primarily involves protection from moisture. Hay stored on the ground will absorb dampness and will mold—on dirt floors, on wood-over-dirt floors, on gravel floors, and sometimes even

Table 11-1. Approximate Space Requirements for Various Feeds

GROUND GRAIN	APPROX. POUNDS/CU. FT.	HAY	CU. FT./TON	POUNDS/CU. FT.
Corn	36	baled	250-330	8-6
Oats	18	baled straw	400-500	5-4
Wheat	43	loose	450-600	4.4-3.3
		loose straw	670-1,000	3-2

on concrete floors. To avoid the mold problem, hay must be stored in lofts or on sleepers at least twelve inches off the ground. Of course, poor-quality hay, no matter how well stored, will still be poor-quality hay. Hay baled too wet will mold (and may pose a threat of spontaneous combustion); hay stored more than a year is likely to be dusty.

When using homegrown grain, be sure it is dry enough before it is bagged up or stored in containers; again, damp grain molds. Purchased grain in sacks should be kept off the floor to circumvent both dampness and rodents. Sweet feeds should not be stored more than a month, if possible. Ideally, most grains should be purchased in quantities that will be used up in three to four weeks. (Sweet feeds are even more perishable in hot weather.)

Metal bins, or extremely tight wooden ones, are good for grain storage. The problem with wood is that it can be chewed through, but wood bins can be lined with metal if rodents are a real problem. A 55-gallon drum is an ideal rodent-proof container for 150–200 pounds of grain. For most smallholders, a few drums will do the trick.

Now it is a fact of life that animals get out of wherever they're supposed to be and end up wandering around the barn, checking out forbidden territory and—here's the important part—looking for goodies. At best, a beast in the grain will merely wreak havoc— spilling it, walking on it, *wasting* it—but at worst, and not uncommonly, a sudden grain feast will make your livestock quite ill. Locating all the grain in one area and making it tight against marauding livestock will eliminate this danger. Keeping grain tightly covered will also serve this purpose, but it is another fact of life that tight lids aren't always so tight. Therefore, leaving feed accessible, although tightly covered, isn't as good a solution. Another reason to concentrate grain in one area is for rodent control.

Rodent Control

I don't know if there's anything you can do, or really need to do, about your run-of-the-mill mouse population. As long as you're not overrun with them, mice don't present much of a problem. (Remember, if you see a good-sized snake in your barn, don't panic. Instead, count your blessings—snakes are great at helping to keep down the mouse population.) Rats, however, are mice of a different color. Rats can do an enormous amount of damage and can carry many diseases, so they must be kept under strict control. With luck, you won't have rats in the first place, making rat control a question of prevention rather than cure, which means, basically, keeping your barn from becoming Rat Heaven. The two major components attracting rats are plenty of food and plenty of shelter. Keep feed in metal bins, either covered and/or at least six inches from any vertical surfaces. Keeping feed in one area, which can be (and is) swept up, reduces the number of places where feed can be found. Where feed is left out for ad lib feeding (of your animals, not of the rats), hanging feeders (for chickens) or feeders with lids (for pigs) will discourage rats. Be particularly aware of signs of rats in these areas. Shelter is the other attraction. Avoid scattered piles of debris, and keep one area for storage rather than leaving clumps here and there. Weeds along the walls of the building also help rats feel secure.

Spotting a rat problem is not difficult: their droppings are easy to see, and a nocturnal trip to the barn with a flashlight (or a quick switching on of the lights) will reveal disappearing tails. (If you're lucky, some rats will quite brazenly go on eating that baby chick right in front of you.) Once you know you have rats, eradication measures should be initiated immediately. First, check your feed setup and clean up piles of debris. Then, the easiest thing to do is to set rat traps (giant mouse traps which can break your fingers quite easily). Baited with a piece of meat (tied on) and a glob of peanut butter, the rat trap always gets its rat. Make sure, though, that the trap is inaccessible to other animals, because the trap, if it doesn't get its rat, will always get *something*. Once you've trapped a few rats and the traps have remained unproductive for a few days, you can assume that the problem is under control—especially if you no longer find any droppings. If, however, the problem is bigger than a few traps can handle, get in touch with your county extension service for assistance in developing an eradication program.

Once your storage vessels are worked out, your rodents under control, and your feed space set up in an area convenient for most of the pens, the only thing left to do is to arrange things so that you can feed efficiently. To measure out feed, you can use scoops (which come in a variety of sizes), or more economically, one- or two-pound coffee cans. Not all feed weighs the same per volume, though, so when you get a new feed, weigh the amount you scoop so you know how much to feed. Locate feeders on the aisle side of the pens; you should never have to go into a pen to feed the inhabitant. If several animals are getting the same feed, devise a system for carrying the full quantity needed so you don't have to keep running back and forth.

Watering

Watering should be set up to be as easy and as fast as feeding. Assuming you do not have automatic waterers, and you are not willing to invest in installing them for just a few head, you'll be using the bucket system for watering individual animals. Water buckets, like feeders, should be located so you do not have to enter the pen. How you water the animals depends greatly on your water supply to the barn. A system commonly used here in the freezing North is a frost-free hydrant in the barn which empties into a barrel or a tank. A bucket is kept hanging by the tank; this is the scoop bucket used to carry the water to the animals' buckets. The trick here is that the scoop bucket is never put down anywhere; that way it stays clean and will not contaminate the tankful of water when it is dipped in, as would each animal's bucket, or as it would if you plopped it down on the floor every time you ran to get some grain or something. An alternative is to bring each animal to the tank twice a day to allow it to drink—a less satisfactory system because of contamination problems and extra labor. Another method is to have a hose that can reach each water bucket—a nice system but may be a problem when freezing weather sets in. (Frozen hose breaks easily, and unless you drain it thoroughly after each use, will quickly block up with ice.)

For watering several animals as a group, an old bathtub is an excellent and inexpensive alternative to a purchased watering trough. When selecting a watering vessel, avoid seams soldered with lead. For watering individuals, large rubber buckets, although initially expensive, are well worth the investment where freezing water is an issue. Breaking ice out of rubber buckets is relatively easy; bounce them off something hard and the ice cracks out. But getting ice out of metal buckets is a pain; if you bounce them they break, and chipping ice is slow and often results in punctures. The same is true for spackle buckets, which often crack before the ice does. With float valves, hydrants, deicers, hoses, etc., there are a million ways you can set up your watering system winter and summer. The main thing is that you plan your system to provide clean, uncontaminated, fresh water easily and efficiently and in sufficient quantities (see Table 11-2).

Table 11-2. Quantities of Water Needed by Farm Animals per Day

SPECIES	GALLONS/DAY
Cow, dairy	10-40 (3-4 lbs. per pound of dry matter consumed plus 3-4 lbs. per pound of milk produced)
Cow, beef	3-12
fattening calves	6-8
Horse, mature	8-12
foals	6-8
Pig, 50-200 lbs.	¾-1½
Sheep, wintering ewes	1
nursing ewes	1½
fattening lambs	½
Chickens, 100 layers	4-11

Manure Handling

Manure disposal has become a major problem for commercial agriculture, and is, at best, a minor one for the smallholder. But it can still be a headache if a system isn't worked out ahead of time. The types of systems available to the smallholder are rather limited—wheelbarrow and pitchfork about covers it (as opposed to the large-scale possibilities of lagoons, oxidation ditches, liquid manure systems, etc.). However, some variation is introduced when considering what to do with it once it's in the wheelbarrow.

When planning the manure system, ask yourself several questions. How much are you dealing with? What type? How much land is available to spread it on? Are you near ponds or streams where runoff pollution might be a problem? What kind of neighborhood are you in: will the neighbors object to a manure heap, or will they be delighted to have a good supply of fertilizer for their gardens?

The easiest and most practical method, where possible, is to stack the manure and then spread it (or have it carted away by gardeners) a couple of times a year. The problems with this system, though, are many: odor, flies, runoff, loss of fertilizer value, and objecting neighbors. Therefore, the location and the type of heap are important. If a covered area within or adjacent to the barn is available, runoff, pollution, and nutrient loss will be minimized. Odors can be reduced by packing the manure tightly and occasionally sprinkling the heap with lime. If you build the heap up during the winter, then spread it during the spring, and don't start another heap till the fall (keeping animals on pasture eliminates manure heaps), problems with flies will also be reduced. An outside heap should be located where it will be least visible to neighbors and will cause no pollution. A shallow, straight-sided pit in which the manure is packed evenly and tightly will lessen unsightliness, runoff, and odors. Flies, however, may still be a problem. Allowing chickens to scratch over the manure heap can help: they peck out some

Table 11-3. Nutrients in One Ton of Manure

GENERAL FORMULA: .5-.25-.5

SPECIES	WATER (%)	NITROGEN (N) (lbs.)	PHOSPHOROUS (P) (lbs.)	POTASH (K) (lbs.)
Sheep	66	21	6	19
Horse	74	13	5	15
Cow	83	10	4	7
Steer	80	15	6	8
Pig	86	10	7	13
Poultry	55	20	16	8

From J. A. Silpher, Ohio State University, Bulletin 262.

larvae and spread and turn the top layer enough to dry it out by exposing it to the sun, making unfavorable conditions for fly breeding and egg hatching.

If you have enough animals to justify the purchase of a manure spreader, get one; the easier it is to spread, the less likely it is to build up. Otherwise, encourage gardeners to come and cart the manure away for you (the easiest, most labor-free method of disposal) and spread the rest by using a truck, wagon, or what have you.

When planning your barn layout and considering the efficiency of feeding, manure handling, etc., don't forget the example set by those old, three-story barns of yesteryear. The barn was built into a hill so a hay wagon could drive right into the top floor. The animals, entering from the lower side, were housed in the middle level. A covered bottom bay held the manure spreader. In this way, hay was unloaded right where it was stored and was simply dropped down a chute to the animals below at feeding time. Cleaning and manure handling also used gravity, not muscle: the manure was scraped or pitched into the aisle, and then scraped down to the bay where it was pushed into the spreader. When the spreader was full, the manure was removed. Many of the old barn plans, although disasters for modern mechanized commercial agriculture, are models of ingenuity and efficiency for the smallholder; let them be a lesson for us all.

References

ENSMINGER, M. E. *The Stockman's Handbook*. Danville, Ill.: The Interstate Printers and Publishers, 1960.

12
FIRST AID

Obviously, the first consideration in first aid is to avoid the need for it, but no matter how careful you are, injuries will occur. Having a first-aid kit on hand, knowing how and when to use it, and knowing when to call the vet will make your ownership of livestock safer and easier for all concerned.

First-Aid Kit

Locate the first-aid box where it can be kept clean and easily accessible. (Storing materials in plastic bags helps keep them clean.) During the winter it may be necessary to keep it in the house or in some other spot where the liquids will not freeze.

A good first-aid box, suitable for most livestock, includes:

thermometer	balling gun
sharp knife	measuring cups and spoons
adhesive tape	hydrogen peroxide
gauze pads	alcohol
gauze bandage	petroleum jelly
cotton	Epsom salts
3-inch-wide bandages	nitrofurazone-based spray and
twitch for horses; halter for cattle	ointment for open wounds, chafes, burns

sponges

Pepto Bismol, milk of magnesia, mineral oil, and colic remedy for horses (as recommended by vet)

liniment

tincture of iodine

dose syringe

antibacterial liquid soap

flashlight

General Procedures

Early detection of injuries and illnesses is one of the most important parts of first aid. Observe your animals daily—really look at and be aware of them—for signs of illness: listlessness, lack of appetite, unusual behavior, wounds, etc. The following are some general rules for dealing with a sick or injured animal:

1. Keep yourself and the animal calm.
2. If possible, put the animal in a clean, draft-free stall (if appropriate) and/or isolate it from other animals.
3. Use whatever restraint may be necessary to keep the animal from injuring itself or its handlers.
4. As much as you can, determine the extent of the illness or injury and try to recognize symptoms of various conditions (i.e., coughing indicates respiratory trouble; fever, an infection; flank biting, intestinal distress; etc.).
5. Know how to use your first-aid equipment.
6. Call for professional help promptly if it is indicated.

When to Call the Vet

Initially you'll probably call the vet more often than is really necessary, which is okay, since it is only through experience that you will learn when to make that phone call. Once you have determined that the problem needs the attention of the vet, call right away; don't wait around to see what happens. When in doubt, call. The vet doesn't want to come out unnecessarily any more than you want him or her to, so be prepared to present as much information about the animal and the situation as possible (including temperature). Sometimes the vet will be able to give you instructions over the phone to tide you over until he or she can get there, or occasionally, instructions on how to treat the animal yourself, with directions to call back after some period of time. Be prepared to carry out the instructions; livestock owners cannot be squeamish.

Generally, you'd want to call a vet for:

1. temperature more than two degrees above normal
2. deep wounds, puncture wounds, or wounds near joints

3. bleeding that cannot be stopped
4. lameness
5. lack of response to treatment
6. excessive, continued nasal discharge, eye discharge, or cough
7. problems during deliveries
8. an animal off feed and showing signs of intestinal distress (lack of bowel passage, eructation, or rumen activity; or pawing, flank biting, rolling, etc.)

Common Situations and Procedures

Temperature, Pulse, and Respiration

Taking a temperature with a veterinary thermometer is the same as with any other rectal thermometer. Shake it down, moisten the bulb, and insert it several inches into the rectum, leaving it in at least three minutes. To avoid possible mishaps, tie a string with a clip to the end of the thermometer, and fasten the clip to the hairs of the animal's tail. (This helps prevent the disappearing thermometer phenomenon.) After use, clean the thermometer with alcohol. Temperatures may be slightly elevated for several reasons: excitement, the heat of the day, or just normal variation.

Pulse (beats per minute) can be taken on any large artery you can find; naturally, locations vary with species. Pulse rates, too, can be elevated for reasons other than illness: excitement, overexertion, heat, etc.

Respiration rate can be measured by watching the animal's flank or by holding a hand in front of the nose, counting breaths per minute.

Table 12-1. Temperature, Heart Rate, and Respiration Rate of Common Farm Animals

ANIMAL	RECTAL TEMPERATURE (°F ± 1°F)	HEART RATE (avg. beats/min.)	Range	RESPIRATION RATE (breaths/min.)
Horse	100.5	44	23–70	12
Cattle	101.5		60–70	30
Sheep	103	75	60–120	19
Goat	104	80	70–135	20–24
Pig	102		58–86	
Rabbit	102.5	205	123–304	39
Adult chicken			250–300	12–20(M) 20–36(F)
Chick			350–450	

Wounds

Wounds fall into four categories: incisions, lacerations, punctures, and contusions; but some statements can be applied to wounds in general.

1. Body wounds heal more quickly than leg wounds.
2. Keep wounds dry; avoid greasy salves or ointments.
3. Do not use extremely strong antiseptics.
4. Steady bleeding indicates a cut vein; spurting blood indicates an artery.
5. Bleeding can usually be stopped by applying a sterile pressure bandage or by direct pressure.
6. A tourniquet is necessary only if an artery has been cut and you cannot stop the bleeding. If one must be used, release it frequently. A novice should try to get advice on the use of a tourniquet, if possible.

More specifically:

Incised wounds are clean, straight-edged cuts. If the cut is small, wash it well with antibacterial soap; then apply wound dressing. Bring the edges of the cut together with a bandage. If the cut is long or quite deep, call the vet, as stitches may be necessary. Until the vet arrives, keep the cut area clean and control bleeding by using direct pressure with clean gauze.

Lacerations are tears with ragged edges (frequently caused by barbed wire); they heal more slowly than incised wounds. Because the risk of infection is very great, usually the vet should be consulted. Keep the cut clean, and control the bleeding.

Punctures, too, are quite prone to infection and should be treated like lacerations. Remove the object causing the puncture (if it's still there), clean the wound, and use a dressing to keep it clean. Consult a vet.

Contusions are the result of broken blood vessels under the skin (the skin itself may or may not be broken), causing swelling and discoloration. If the skin is cut, treat accordingly. If the skin is not cut, apply ice packs to the bruise three to four times over the first twenty-four hours (or run a hose on it) to reduce the swelling. Once the swelling has gone down, moist heat from hot towels or hot water will help promote circulation in the area.

Lameness

Lameness associated with a cut or bruise can be handled as above, but otherwise it can be a sign of a more serious illness or disease. A slight lameness should clear up within a few days (use liniment), but if it doesn't, or if it gets worse, call the vet. If the lameness is associated with fever, listlessness, or any other signs of illness, again, call your vet.

Digestive Disorders

Digestive disorders come in many shapes and sizes, and the type you'll be dealing with depends on the species of animal. Bloat, scours, and indigestion are discussed in the chapters on ruminants; colic in Chapter 10.

The basic sign of digestive trouble is the animal going off feed and/or a change in the frequency or consistency of the bowel movements. Except in clear cases of simple indigestion, it is best to consult your vet.

Respiratory Ailments

Persistent cough, nasal discharge, and possibly a slight fever are indicative of colds. Keep the animal quiet, dry, and out of drafts. Dampening dry or dusty grain and hay will lessen the aggravation of the cough. A cold should clear up in a day or two. If it doesn't, if the temperature goes up, or if the discharge becomes thick and yellowish, call the vet—untreated colds can turn into pneumonia. High fever, rapid breathing, and weakness are additional signs of pneumonia; call the vet immediately.

Giving Injections

It may come to pass that you will have to give an injection. Rarely would you be required to give an intravenous injection (in the vein), but it is well within the realm of possibility that you'll have to give intramuscular (in the muscle) or subcutaneous (under the skin) ones. The easiest and best way to learn to give injections is to have your vet or an experienced friend show you how. Barring that, however, the basic steps are as follows:

1. Use a sterile syringe and needle of the proper size for the animal (18 to 21 gauge, by 1 to 1½ inches long for larger animals; 21 to 25 gauge for smaller) and for the amount of medication being given (5 to 20 cc syringes are most common).

2. Shake the bottle of medication (usually; read the directions first), hold it upside down, insert the needle, and withdraw the specified amount. Pull out the needle, hold it upright, and tap the syringe to move air bubbles to the tip. Depress the plunger slightly to get rid of the bubbles.

3. Restrain the animal adequately.

4. Clean the injection site with alcohol.

5. Intramuscular injections are given into large muscle areas (neck, thigh, etc.). Insert the needle swiftly, approximately ½ inch, perpendicular to the animal's body. Pull back on the plunger

slightly to check for blood; if there are any signs of having entered
a blood vessel, find a new spot. Then inject the medication, withdraw
the needle, and wipe the area with alcohol. A quick note here on
plunging the needle into the animal's skin. This seems to be the
most difficult moment for first timers; perhaps the sense of identification is too great. At any rate, the faster and smoother you can
get the needle in, the better for all; so just be courageous and try
to get it in on the first stab. One thing that helps is to have someone
distract the animal at the critical moment, like petting one side of
the neck while you jab the needle swiftly and dextrously into the
other side. Another trick is to hold the syringe between your thumb
and index finger with the needle pointed toward your pinky. Tap
the injection site with the edge of that hand a few times; then, on
one of the taps, stick in the needle. Another point to remember is
that animal skin is tough, and a certain amount of force is necessary
to insert the needle (horses are an exception).

   6. *For a subcutaneous injection,* find an area of loose skin on
the animal. Pinch a section and slip the needle into the fold, holding
the needle parallel to the animal's body. You can feel the needle
inside the skin. Inject the medication, withdraw the needle
smoothly, and wipe the area with alcohol.

   7. *When injecting large quantities* or making repeated injections, alternate sites.

   8. *Keep calm.* If the animal struggles while the needle is in,
withdraw or let go of the syringe, let the animal calm down, and
try again.

   9. *Store* unused medication as directed on the bottle.

Giving Other Medication: Pills, Powders, Liquids

   Clearly, the trick of giving a pill is to get the animal to swallow
it. Trickery and deceit sometimes work; I hid a pill inside an apple
once and fed it to my horse. It worked great for one pill, but it was
three days before the horse would eat another apple, so new tricks
were needed for the second pill. Sometimes you can pulverize a pill
and treat it like a powder; add it to the grain with a little molasses
and it will usually be eaten. (For animals who can sniff out powders
or pills a mile away, a little Vicks VapoRub on the nose confuses
the senses just enough.) Occasionally an animal will shock you by
being willing to eat the thing right out of your hand. (Always try
that first: "Look at the treat I've got for you!" Beware of spit out,
rejected treats, though.) Once trickery has failed, a measure of force
is necessary. If possible, force the animal's mouth open, deposit the
pill as far back as you can, hold the mouth closed, and stroke the

neck until the beast swallows. However, watch out for those who swallow, look at you innocently, swallow again obligingly and loudly, and then, as soon as you turn your back, spit the pill out into a pile of manure.

A more commonly used method is the balling gun. A balling gun holds the pill and is inserted well into the side of the mouth; the plunger is plunged, and the pill is released. It's actually quite simple.

Drenching (giving liquids) is harder because it is not so well suited to trickery as is giving pills, and you must be careful to avoid pouring the liquid down the windpipe. Holding the head at a gentle, rather than severely upright, angle will help avoid this problem. A lamb nipple with an enlarged hole attached to a heavy-duty bottle works well, as does a plastic bulb baster. Pour liquid in slowly from the side of the mouth.

As you become more experienced with first aid, you will be able to treat many of the routine injuries and simple illnesses yourself. But remember, there is no substitute for good, prompt treatment, so don't become overconfident; always be ready to discuss the problems with your vet.

References

First Aid

Animal Diseases. 1956 Yearbook of Agriculture, Washington, D.C.: U.S. Government Printing Office.

BODDIE, G. *Diagnostic Methods in Veterinary Medicine.* Philadelphia, Pa.: J. B. Lippincott Co., 1969.

BRANDER, G. C., and D. M. PUGH. *Veterinary Applied Pharmacology and Therapeutics.* Baltimore, Md.: William Wilkins Co., 1971.

FRASER, A. F. *Farm Animal Behavior.* Baltimore, Md.: William Wilkins Co., 1974.

Merck Veterinary Manual. Rahway, N.J.: Merck and Co., Inc., 1973.

SPAULDING, C. E. *A Veterinary Guide for Animal Owners.* Emmaus, Pa.: Rodale Press, Inc., 1976.

SWENSON, M. J., ed. *Duke's Physiology of Domestic Animals.* Ithaca, N.Y.: Cornell University Press, 1970.

Supplies

ANIMAL MEDIA INC. Livestock Supplies. Manchester, Pa. 17345

CREUTZBURG INC. Livestock Supplies. Malvern, Pa. 19355

OMAHA VACCINE Co. 2900 "O" Street, Omaha, Neb. 68107

KANSAS CITY VACCINE Co. Farmade Products Handbook. Stockyards, Kansas City, Mo. 64102

Appendix I

Using Feeding Standards

For many years, Morrison's feeding standards (see "References") were the ones prevalently used, but recent research has evolved more accurate standards, which are published by the National Research Council on Nutrition (NRC). However, in spite of the fact that they are somewhat dated, I find Morrison's easier to use; also, Morrison's *Feeds and Feeding,* which contains the standards, is a valuable reference book for the smallholder and is easy to come by as a used college text. If the standards are taken as a guide, rather than as gospel, Morrison's tables should be adequate. If you are formulating complicated rations on a large scale, it is worthwhile to get hold of the NRC standards.

The example given uses Morrison's tables for the animals' requirements and for the composition of feeds, but the technique is applicable to the NRC tables as well.

The problem: You have a 1,000-pound dairy cow in her sixth month of lactation and fourth month of pregnancy, giving forty pounds of milk per day with 4 percent fat. You're lucky and have good-quality alfalfa hay to feed. You can buy corn from a local farmer or a commercial mix.

The solution: There are two routes you can choose between. The first:

1. Figure out how much hay the cow is consuming, assuming two pounds of hay per one hundred pounds of body weight.

2. Using Morrison's Table VII, determine how much protein is needed in the concentrate mix, based on the roughage fed. In this case, it is 12 percent.

3. Using a grain-feeding guide (such as Table 3-2), knowing the pounds of milk produced and the hay consumption, determine the pounds of concentrate to be fed. A 40-pound-per-day milker, getting 20 pounds of hay per day, with 4 percent milk, needs 13.7 pounds of concentrates per day.

4. Go buy a commercially mixed 12 percent protein feed, and feed it to the cow at that rate; or develop a 12 percent protein feed of your own, using the available corn with a purchased soy meal supplement. To formulate a 12 percent mix, you can use the Pearson's square method:

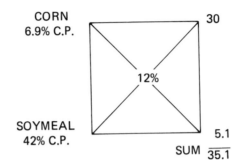

CORN
6.9% C.P.

30

12%

SOYMEAL
42% C.P.

5.1

SUM 35.1

a. Put the desired protein level in the center of the square,
b. Find the crude protein level of each of the feeds, and put them on the left of the square as indicated.
c. Take the difference between the crude protein level of the feed ingredient and the final protein level desired (in the center), and enter it on the right diagonal across the square (6.9 − 12 = 5.1; 42 − 12 = 30).
d. Add the numbers on the right (30 + 5.1 = 35.1).
e. To find the percentage of each ingredient in the final ration mix, divide the number opposite the ingredient, on the right, by the sum.

$$\text{percent of corn in the ration: } \frac{30}{35.1} = 85.5$$

$$\text{percent of soy meal in the ration: } \frac{5.1}{35.1} = 14.5$$

A 100-pound bag of this feed would contain, therefore, 85.5 pounds of corn and 14.5 pounds of soy meal, and it would be a 12 percent protein mix.

5. *When formulating your own mix,* you must pay attention to vitamins and minerals. Trace mineralized salt should be provided, but use the standards to establish what more, if anything, might be needed (see below). A commercially mixed complete feed will usually have the vitamins and minerals in sufficient quantities.

A second route may be taken:

1. *Establish the daily nutrient requirements* of the animal, which are as follows:

	Lbs. D.P.	Lbs. TDN	Lbs. Ca	Lbs. Phos.	Mg. Carotene
Maintenance	.60-.65	7.0-7.9	.018	.018	40
Additional per lb. of milk	.045-.057	.31-.32	.0022	.0017	—
Total for milk production (40 lbs./day)	1.8-2.28	12.40-12.80	.088	.068	—
Total daily requirement, maintenance plus milk	2.40-2.93	19.40-20.7	.106	.086	40

2. Establish the amount of nutrients derived from the roughage being fed (alfalfa hay) from tables of feed composition:

	DRY MATTER	D.P.	TDN	Ca	Phos.
		(percent)			
Alfalfa hay	90.5	10.9	50.7	1.47	.24

If the cow eats twenty pounds of this hay per day, the amount of nutrients she receives is

D.P.	TDN	Ca	Phos.
	(pounds)		
2.18	10.14	.294	.048

This figure is calculated by multiplying the amount eaten times the percent nutrient in the feed:

20 lbs. $\times$ 10.9 = 2.18 lbs. D.P.
20 lbs. $\times$ 50.7 = 10.14 lbs. TDN

3. The difference between what is supplied in the roughage and what the animal requires must be supplied by the concentrate.

	D.P.	TDN	Ca	Phos.
Requirements	2.40-2.93	19.40-20.7	.106	.086
Supplied in roughage	2.18	10.14	.294	.048
Balance required	.22-.75	9.26-10.56	ample	.038

As you may have guessed, the alfalfa hay provides the bulk of the digestible protein required and more than enough calcium. TDN and phosphorous are low and must be supplied by the concentrate.

4. Since corn is available, find the nutrients supplied by corn from the feed composition tables.

	D.P.	TDN	Phos.
	(percent)		
Corn, dent, grade 2	6.7	80.1	.27

5. *From the grain-feeding guides, estimate* the amount of concentrate to be fed, and calculate the nutrients supplied.

AMOUNT ESTIMATED	COMPOSITION	NUTRIENTS SUPPLIED
12 lbs. corn	× 6.7% D.P.	= .80 lbs. D.P.
12 lbs. corn	× 80.1% TDN	= 9.6 lbs. TDN
12 lbs. corn	× .27% Phos.	= .032 lbs. Phos.

6. *Comparing the nutrients* supplied by the corn with the balance required (step 3), you can see that the digestible protein requirement is well met, the TDN is adequate, and the phosphorous

Table I-1. Proximate Composition and Digestible Nutrients of Some Common Feeds*

	DRY MATTER (%)	PROTEIN (%)	FAT (%)	FIBER (%)	NITROGEN-FREE EXTRACT (%)	ASH (%)	DIG. PROTEIN (%)	TDN (%)
Hays								
Alfalfa hay, all analyses	90.5	15.3	1.9	28.6	36.7	8.0	10.9	50.7
Alfalfa hay, leafy	90.5	16.0	2.1	27.2	36.8	8.4	11.7	51.2
Alfalfa hay, fair	90.5	13.7	1.7	31.8	36.0	7.3	9.7	50.3
Timothy hay, before bloom	89.0	9.7	2.7	27.9	42.2	6.5	6.1	56.6
Timothy hay, late seed	89.0	5.3	2.3	31.2	45.7	4.5	1.9	41.9
Pastures								
Orchardgrass, pasture	23.9	4.4	1.2	5.6	10.0	2.7	3.2	15.9
Timothy, pasture stage	23.9	4.7	0.9	4.6	11.1	2.6	3.5	16.5
Silages								
Corn silage, well matured	27.6	2.3	0.8	6.7	16.2	1.6	1.2	18.3
Grass silage, small prop'n legumes	27.6	3.2	1.1	9.7	11.1	2.5	1.9	15.6
Concentrates								
Corn #2 dent	85.0	8.7	3.9	2.0	69.2	1.2	6.7	80.1
Oats, not incl. P. C. states	90.2	12.0	4.6	11.0	58.6	4.0	9.4	70.1
Wheat bran	90.1	16.4	4.5	10.0	53.1	6.1	13.3	66.9
Soybean oil, meal solvent process	90.4	45.7	1.3	5.9	31.4	6.1	42.0	78.1

* Data from Morrison's *Feeds and Feeding*, 22nd Edition.

is a shade low. At this point you can declare yourself satisfied with the ration or increase the corn a bit to bring up the TDN and phosphorous.

	D.P.	TDN	PHOS.
		(pounds)	
13 lbs. of corn supplies:	.87	10.4	.035

Depending on the different feed ingredients available, their cost and suitability, the rations can be played around with and altered in many ways. More often, combinations of grains (instead of just one as shown in the example) are needed to meet the requirements most economically. Mineral supplements might also be necessary (dicalcium phosphate or a mineral mix).

The ration evolved by the first route was 20 pounds of alfalfa hay, plus 13.7 pounds of a 12 percent concentrate mix of corn and soy. Via the second route, the ration is 20 pounds of alfalfa hay plus 12 pounds of corn. The rations are not dramatically different, but let's check the ration derived from the guides for protein and TDN.

Concentrate mix: Corn 85.5%
    Soy 14.5%
of 13.7 pounds of mix, that is 11.7 pounds of
corn and 2 pounds of soy meal.

TDN
11.7 lbs. corn × 80.1% TDN = 9.4 lbs. TDN
  2 lbs. soy × 78.1% TDN = 1.6 lbs. TDN
Total TDN from mix   = 11.0 lbs. TDN

Protein
11.7 lbs. corn × 6.7% D.P.   = .78 lbs. D.P.
  2 lbs. soy × 42% D.P.   = .84 lbs. D.P.
Total digestible protein from mix = 1.62 lbs. D.P.

Feeding this corn-soy mix at a rate of 13.7 pounds per day overfeeds protein and TDN slightly; the ration should be adjusted to reduce the protein if protein is expensive. Depending on the relative prices of corn and soy, either the mix or an all-corn ration can be fed.

A final note on the grain-feeding guides. The TDN amounts specified as required by lactating cows in the Morrison standards are low. Because digestibility of feeds decreases as intake increases, for high-producing cows, more TDN is required.

Appendix II

Hay, Pasture, and Silage

You've heard about the grass being greener on the other side. Well, if it is, there might be something you can do about it.

Most people think that any patch of grass constitutes a pasture, and the thicker and taller the growth on it, the better it is. Further, it's assumed that if you've got a wild and woolly patch of field, all you've got to do is throw a fence around it, stick in the animals, and they'll eat it all up. Wrong, from the word go. I am no pasture expert, but here is a simple rundown on types of pastures and how to maintain them.

Land that is too steep, too rocky, or too rough to cultivate (sounds familiar, right?) is usually kept in *permanent pasture*. This is land that is not cropped but is kept in plants that reseed themselves. Common permanent pastures on the more fertile soils are Kentucky bluegrass and white clover, with some redtop, timothy, orchardgrass, and fescues.

Renovated pasture is permanent pasture that has been partially tilled (usually disked) and reseeded to a rapidly growing, large, vigorous species. The field can also be limed and fertilized, and some form of weed control may be used (often herbicides). Among the grasses easily established for permanent pasture (when renovating) are Italian ryegrass, perennial ryegrass, oatgrass, orchardgrass, bromegrass, timothy, and redtop. Among the legumes used are red clover, sweet clover, alfalfa, alsike clover, Ladino clover, and European birdsfoot trefoil.

Semipermanent pastures are usually completely tilled and seeded to some species which will last two to ten years before reseeding. They are usually planted to perennial grasses and legumes, including birdsfoot trefoil, timothy, alfalfa, Ladino clover, bromegrass, and orchardgrass.

Rotational pastures are good, plowable lands which are pastured for one to several years but are included in the rotation scheme (i.e., pasture one year, corn the next, etc.).

During midsummer, permanent pastures decline in quality. To offset this drop, some farmers provide *supplementary pastures*: pastures that come in late and are used to maintain constant forage production throughout the season. Sudangrass and ryegrass are frequently used species.

When hay, silage, or any crop is harvested, either the residue (corn stalks, wheat straw, etc.) or the regrowth of the crop can be grazed as *aftermath pasture.*

A well-maintained renovated pasture can support as much as one cow plus calf per acre. But what is a well-maintained pasture?

To start with, something of the growth cycles of pastures should be understood. A steak may be a steak, but grass is not grass, and at different times during the season, pastures have very different yields and nutritional values. Fig. II-1 shows a curve describing the yields of a Kentucky bluegrass pasture during our local (New York State) season. Obviously, local conditions vary, so check with your extension office about seasonal pasture quality in your area.

Around May and June, pastures will be yielding at their peak: by midsummer, yields will drop. The early spring growth is higher in protein, too; as grass gets older (and taller) it becomes largely indigestible and lower in protein, so its nutritional value drops radically. Mixing pasture species and using supplementary pastures will help solve this problem. Proper management, too, will increase the value of your pastures, regardless of their type.

Remember that young growth is much more palatable and nutritious than long, rank, older growth. A neighbor of ours who hadn't had any experience with livestock fenced in a large field for some beef calves he wanted to raise. The field was chock full of vegetation, but the calves kept breaking out. Why? Not just because the grass was greener on the other side, but because it was shorter, growing, delicious, and nutritious. *Quantity* does not mean *quality.* In fact, when a hayfield, for example, reaches its maximum output, it has already passed its nutritional prime.

To keep your grass from getting overly mature, you can do two things: mow it and/or stock it at the proper rate. If you don't have enough animals to keep the grass short and growing, you must mow it before it heads out. Usually once or twice during the season will be adequate. *Look* at your fields and look at your animals on the

Fig. II-1. Yield of a Kentucky Bluegrass Pasture over a New York State Pasture Season.

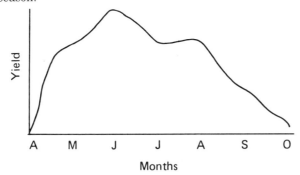

pasture. Are they eating? What are they leaving? Why? Accurate observation and common sense make all the difference in this, as in all phases of management.

Stocking rate is one of the biggest problems in managing your pasture. You can use the guides of one acre per cow and two acres per horse, but much depends on the type and quality of the pasture. Again, using your eyes and your head will help in determining the number of animals you can keep on a field. Remember, though, that because of the seasonal productivity of pastures, you can stock more heavily early in the season but must reduce the stocking rate as the summer progresses.

Another management technique to make the most of your pasture is dividing it. Electric fence is easily used for this purpose. Animals are selective grazers, which means that given the choice, they'll eat the best part of the grass and leave the rest behind. Well, that "rest" can be perfectly decent stuff, now going to waste. By dividing your field into smaller sections and fencing the beasts into one section of it, you'll encourage them to eat the whole plant. Once that section is well grazed (but before it's overgrazed), move them into the next section. This method not only makes the best use of your field but also helps in parasite control. By taking the animals out of one section for several weeks, many of the common parasites' life cycles in that area are broken. Then, when you put the animals back into it, that section of pasture will be relatively clean.

Grazing more than one species together on a field is another age-old, but now ignored, practice, which also helps greatly with parasites. Parasites are fairly specific; usually they work well only in one species. When the eggs of a cow parasite are ingested by a horse, they get wiped out, leaving fewer cow parasite eggs for the cow to take in. Also, different species graze differently. Cows eat grass by wrapping their tongues around it, then pulling (so they take off the longer grass). Sheep or horses, however, can bite off the short grass, so they can come through and clean up what the cow has left. Sheep are good with cattle, too, because in addition to eating different parts of the same plant, they'll eat entire plants that cattle refuse. You'll also notice that animals will not graze immediately around their own droppings, which causes rings of brilliant green grass around each pile of manure. However, a cow will eat closer to horse droppings than a horse will, and vice versa, again resulting in better utilization of the pasture.

Actually, the issue of droppings brings up another point. Because of this phenomenon of not eating around the droppings, even with mixed species you should spread the droppings a couple of times during the year. If you don't have one of those flexible harrows, a chain pulled over the field will do the trick. In addition to encouraging the animals to eat over a greater area, obviously it also fertilizes the field more evenly.

Putting the stock on and pulling them off at the right times also affects your pasture growth. Putting the animals on too early in the spring while the ground is still waterlogged will cause holes where the beasts step, preventing grass growth. Spring grazing should not start until the growth is sufficient and the ground firm enough to withstand trampling. Take the animals off the field in the fall while there's still two inches or so of grass—which will not only help with the plants' ability to winter over but also will make the pasture ready earlier in the spring.

Fertilizing and liming your pastures can significantly improve them. If some legumes are already present in the pasture, adding lime, phosphorous, potash, and manure will encourage them, thereby increasing its nutritive value. Adding nitrogen increases the grass yield and will help get a grass pasture off to an earlier start. On a mixed grass and legume pasture, encouraging the legumes will also encourage the grasses because of the legumes' ability to add nitrogen to the soil.

One final word of caution. In the early spring, the grass is greener *everywhere,* including your side of the fence. Many animals, when turned out for the first mouthful of the season on that fresh, young, early spring green grass, have a tendency to gobble it up and are then susceptible to various digestive problems like bloat and colic. So, no matter how eager you are to free those animals from their winter doldrums and get them out of the barn and onto that gorgeous grass, do it gradually.

Above all, remember that good pastures are often the key to economical small-scale stock raising.

Sheep, goats, horses, cattle—even swine to some extent—all eat roughage. In almost all parts of the United States, because of weather conditions, at least part of the year the roughage needs of livestock must be met with stored or harvested forages, most commonly silage and/or hay. For the smallholder, silage is impractical, and hay is the roughage used whenever pasture is not available. Depending on your location and resources, pasture may never be available, may be available three or four months out of the year, or if you are lucky, as much as eight months a year. In any case, a significant portion of the animals' nutritional needs are going to be met with hay, so you should be able to accurately judge their needs and the types and qualities of hay that will meet them.

Hay

Hay is characterized by the type of plant it comes from: grass, legume, and mixtures of the two. Common grass hays include timothy, brome, oat, bermuda, wheatgrass, and orchardgrass. Legumes include alfalfa, trefoil, the clovers, soybean, and lespedeza. In general, grasses are higher in dry matter and fiber than are legumes, whereas legumes are richer in protein, calcium, and vitamins. Because of these higher protein levels, legume hay is more valuable. However, the quality of the hay, although initially dependent on the plant species, is finally dependent on the harvesting: the plant's stage of maturity when harvested and harvesting and storing conditions. Regardless of the plant species, good-quality hay is early cut and green, has a high proportion of leaves to stems, is not musty or moldy, and contains little foreign matter (e.g., weeds).

The stage of maturity at cutting greatly affects both the palatability of the hay and its nutritional value. As plants mature, their dry matter content increases, as does their lignin (undigestible fiber) content; protein, vitamin, and mineral levels all decrease. Therefore, if all other things are equal, early cut hay will have a much greater feeding value than will late cut hay.

The color of hay is indicative of its carotene content, or Vitamin A value. Long exposure to the sun or excessive fermentation losses (from storing too moist) will greatly reduce the carotene levels in the hay, and it will appear bleached or brown. Good green hay is considered one of the best sources of Vitamin A for livestock; every effort should be made to buy hay green, since carotene losses will continue during storage, even under the best conditions.

Most of the protein, minerals, vitamins, and digestible fiber are found in the leaves of hay plants, making leafiness another important factor in quality. If the hay is harvested too dry, the leaves, which dry fastest, will shatter and fall off, leaving a much poorer quality product.

Storing hay before it is sufficiently dry will result in severe nutrient losses and can present quite a fire hazard. Fermentation will cause the hay to heat up; if the heating is great, the losses of carbohydrates and vitamins will be great, and worse, the hay can combust spontaneously. If the hay is well cured and stored at the proper moisture content, some heating and fermentation will still take place, but the losses will be small. Too-wet hay will also mold, making it worse than useless for feeding to some stock (horses) and poor feeding at best for others (cattle).

Whether the hay has been rained on during curing can also affect its quality. If the hay is cut and in the swath and is then rained on heavily, much of the nutrient value will be leached from it. If the hay is rained on soon after cutting, however, losses are not significant.

What about shopping for hay? All those stories you've heard about never trusting a hay dealer are, if not actually true, based on reality. The person selling hay wants to sell the hay, and let the buyer beware. The seller certainly isn't going to stand there and say, "Well, bad luck, it got rained on hard, so I had to bale it a little wet; it'll be all moldy, so how about a dollar a bale?" Forget it.

You're going to have to shop for hay with some idea of what to look for and an expectation of getting burned at least a few times while you're learning. If at all possible, avoid buying hay sight unseen (over the telephone, for example). If you do, expect the worst.

First, how much hay do you need? Most bales weigh between forty and sixty pounds, so if you need two tons for your cow for the winter, that's seventy to one hundred bales. Weigh a bale or two if you can, so you know what you're dealing with. (Always figure a little extra on your needs, though, to cover losses and emergencies. For example, whereas we have a four- or five-month pasture season in the Northeast, I generally take the pessimistic view that winter is going to last at least nine or ten months, and I am sure to have enough hay in case of that not unlikely event. One year it really paid off: winter wasn't quite ten months long, but my mare broke a leg and had to be fed hay well into the pasture season. Hay was a scarce and expensive commodity that spring, and if I hadn't stock-piled earlier, I would have been up the creek.)

After figuring quantity, decide what quality you'll need (see Chapter 1). Dairy animals and young stock need good, high-protein hays, so you'll want one with at least 50 percent legumes. Although animals late in gestation also need high-quality, protein-rich hay, during early gestation a hay lower in protein (a grass hay) may be adequate. Different species of livestock also have different requirements, so the type of hay should be matched to the animal being fed. If you're dealing with enough animals, it may pay to buy different types and qualities of hay for feeding at different periods to the different stock.

Once you've established what and how much hay you need, ask around about the price. Prices are usually quoted by the ton or by the bale. A per ton price is pretty straightforward and easy to deal with. A per bale price, however, is tricky, because it all depends on the size of the bale. Weigh a couple of representative bales, and translate the per bale quote into a per ton price to see if it is commensurate with the prices you've been hearing. When hay is sold by the ton in large quantities, you have the right to expect a weigh ticket. For small quantities, establish the number of bales per ton by weighing a few bales. Try to buy all your hay during hay season, when stocks are high and prices low. Unless it's a deal you can't pass up (and the stuff looks good) avoid buying hay over six months old.

Look at the color of the hay, remembering that rain will leach the color, as will overexposure to the sun, reducing the vitamin and possibly the carbohydrate value. However, hay that is bleached on the outside may be quite nice and green on the inside; it just means that the outside of the bale has been exposed to the sun, which is not very significant. How do you know, though, unless you break open a bale? This is just what you have to do. If the seller won't let you break one open or says, "There's an open one over there," be wary. Look at the inside color and feel the interior of the bale for dampness or heat. Check the hay (on the inside of the bale) for leafiness, for the proportion of legume to grass, and for weeds. Look for whiteness (from overheating) and mold (brown or blue fuzzies on the stems). Smell it; it should smell fresh and sweet. Moldy or musty hay *smells* moldy; trust your nose rather than your eyes if you can smell mold but can't see it. Dustiness can be checked on unopened bales by dropping the bale on a concrete slab and watching for billowing dust clouds.

Upon occasion you may come across standing hay for sale. In this situation, you buy the field, so to speak, and it is baled either by the owner, yourself, or some third party. Buying standing hay can be quite difficult and complicated: terms must be clearly established to insure yields, quality, delivery, etc., plus you must know plant species, weeds, stages of maturity, and so on, in order to judge the value of the field. Although this can be an economical way to buy hay, it is not something to be attempted lightly by the novice.

Silage

Silage is one of those things that most people unconnected with livestock never even heard of. Silos they know—those picturesque towers one sees in typical farm scenes. But what's inside of them? Silage is also one of those things that is not particularly practical for the small farmer: the cost of equipment, unless you already have it all, is prohibitive. Also, making silage in the small quantities needed for just a few head is not generally workable. Nevertheless, anyone attempting to learn about livestock should know what's in those silos and how it is used, if only because silage feeding is so widely practiced in the livestock industries throughout the country.

In twenty-five words or less, silage is a forage made by cutting a crop green, packing it into an airless container, and allowing it to ferment in order to preserve it. The most commonly ensiled crop is corn; the entire plant, ears and stalks, is chopped and put in the silo. One big advantage of making a corn crop into silage rather than corn grain is that more TDN is being harvested per acre. Silage is also easy to handle, store, and feed with mechanical systems. However, although corn silage is an excellent, high-energy feed, it is low in protein, calcium, phosphorous, and possibly some trace minerals; proper supplementation, therefore, is important with corn silage rations.

Hay crop silage is made from (you guessed it) hay crops; the same forages that could be harvested as dry hay (legumes, grasses, or mixes) are instead made into silage. The advantages of hay crop silage (haylage) over dry hay are many: nutrient losses are lower, the labor requirement is less, haylage adapts well to mechanical feeding systems, and the animals eat it almost without waste. Also, crops that would have made poor hay can still make good silage. An advantage on farms where corn silage is made is that it and haylage require almost the same machinery, as opposed to the entirely different set of machinery needed for dry hay. But speaking as an inhabitant of the second cloudiest region in the United States, a huge advantage of haylage over dry hay is that in harvesting it you are not completely at the mercy of the weather. To be free of that hideous anxiety, "To cut or not to cut?" when the fields are ready and the sky is *still* cloudy, is an unimaginable joy, at least to this haymaker. Haylage does have its disadvantages, though: it smells awful; animals may have to be coaxed into eating it initially (although they soon learn to like it); it is subject to freezing; and because of the low sugar content of the crop, it can be tricky to make well. Compared to corn silage, hay crop silage is higher in protein, calcium, and phosphorous but lower in TDN.

How does the ensiling process work? Green forage is cut and chopped short, put into the silo, and packed down. For a while, the plant cells continue to breathe, using up the oxygen and giving off carbon dioxide. This process goes on for about five hours, at the end of which all the air is used up, which prevents mold formation (mold cannot grow in the absence of oxygen). Then, acid-producing bacteria multiply in the silage, "eating" the sugars in the forage and producing, primarily, lactic acid. The acid production is critical, because the acidity prevents the growth of other, undesirable bacteria, such as those that would cause rot. After a certain amount of acid has been formed, the fermentation slows down, and after a few weeks, it just about stops. As long as no air gets into it, the silage will now keep for quite long periods of time. If air does get in, molds will grow and destroy the acid, paving the way for the bacteria that can cause the silage to spoil. The moisture and sugar content of the crops ensiled affect the process: too wet, and an undesirable type of fermentation may take place; too dry, molds might grow; too little sugar, not enough lactic acid is formed.

Silage can be stored in anything from plastic bags to those big, airtight, glass-lined "blue angels" you see on the larger, more prosperous farms. Three types of silos most often used are bunkers (concrete walled pits, which after filling, are often covered with plastic to cut down on surface spoilage), concrete towers, and airtight towers. All of these types contain quantities of silage ranging from 500 to 4,000 or more tons.

The small quantities needed by a smallholder present a whole different set of problems. Various small-scale possibilities include (1) a wood-lined trench, a tractor being driven over the material to pack it as it's put in, it then being covered with heavy, weighted plastic. (2) Small aboveground silos made from metal forms are lined with polyethylene tubing; a commercial vacuum cleaner sucks out the air after the silage has been packed in tightly; the ends of the bag are then folded and tied to seal it. (3) Wire forms are lined with heavy roofing or other waterproof paper, then sealed on top with dirt or plastic. Whatever the form of the silo, the silage must be chopped, and the container must be as airtight as possible, waterproof, and able to withstand the pressure of the tight packing. Of course, once you open the silo to feed out of it, the surface is exposed to the air, so you want to make the system such that the area exposed each time is minimized. (Generally speaking, you want the diameter of the silo to be such that 1½ to 2 inches of silage is removed from the entire surface for feeding daily. Two inches of corn silage weighs about 8 pounds per surface square foot in a silo filled to 30 feet, so a silo 10 feet in diameter would require a feeding level of about 630 pounds per day.) Freezing along the side of the silo also presents a problem in small silos. Then, too, for the smallholder, the problem of chopping the forage and packing it down adequately is not a minor one. Silage has so many advantages, though, that if only someone could come up with a system for small quantities it could make a big difference for the small farmer.

What kind of quantities are needed for feeding? In very general terms, per day, cows in milk will consume about 30 to 70 pounds, beef cows 30 to 50 pounds, fattening steers 25 to 30 pounds, sheep 2 pounds per 100 pounds of body weight, lambs 1½ to 3 pounds. These figures are, to say the least, vague, but they are in the ballpark. If you have one cow, then, for a winter feeding period of seven months, feeding at 50 pounds per day you'd need about 5 tons of silage (see Table II-1).

More specifically, silage feeding can be expressed in hay equivalents and worked out in rations in the usual way (see Chapter 1).

Table II-1. Silage Consumption per Day

SPECIES	POUNDS
Dairy cattle, per 1,000 lbs. body weight	40
Beef cattle, per 1,000 lbs. body weight	30
Horses, per 1,000 lbs. body weight	20
Sheep	2

References

KIPPS, M. S. *Production of Field Crops.* New York: McGraw-Hill Book Co., 1970.

MORRISON, F. B. *Feeds and Feeding,* 9th ed. Claremont, Can.: The Morrison Publishing Co., 1961.

Appendix III

Livestock Data Summary

Table III-1. Fact Sheets

	Beef Cattle	Swine	Dairy Cattle	Sheep	Goats	Horses
Birth weight (lbs.)	50-100	2-3	50-100	5-15	5-15	50-100
Weaning age	175-240 days	3-6 weeks	2 mo.	3-6 mo.	2 mo.	6 mo.
Weaning weight (lbs.)	350-600	30-35 (6 wks)	—	60-100	—	—
Mature weight (lbs.) ♀	1,000-1,700	400-700	900-1,600	100-250	100-135	750-1,500
♂	1,600-2,500	500-800	1,400-3,000	150-300	135-175	750-1,900
Estrous cycle (days)	19-21	21	19-21	16-17	21	21
Duration of estrus	13-17 hrs	2-3 days	13-17 hrs	30-36 hrs	2-3 days	5-7 days
Gestation length (days)	280-283	112-114	280-283	145-150	149-154	336
Number born (avg.)	1	7-15	1	1-3	1-3	1
Breeding age	600-700 lbs. yearlings to calves as 2-yr.-olds	8 mos.	14-19 mos.	1 yr.	8 mos.	2-3 yrs.
Breeding season	spring-summer	late fall/spring	all	fall	fall	Apr.-Oct.
Normal season of birth	spring	spring/fall	all	spring	spring	spring-summer
Slaughter weight (lbs.) ♀	750-950	190-240 at 5-7 mos.	—	90-110	—	—
♂	900-1,100					
Fleece weight (lbs.)	—	—	—	6-14	—	—
Names of classes						
females before 1st birth	heifer	gilt	heifer	ewe	doeling	mare
female	cow	sow	cow	ewe	doe	mare
male	bull	boar	bull	ram	buck	stallion
castrated male	steer	barrow	steer	wether	—	gelding
castrated mature ♂	stag	—	stag	—	—	—
young female	—	—	—	—	—	filly
young male	—	—	—	—	—	colt

311

INDEX

Absorption of nutrients, *(table)* 13
Animal breeding, 20-34 (see Genetics)
Artificial insemination, 65 (see specific species

Barn, 282-289
  design, 282
  manure handling, 288
  planning, 283
  rodent control, 285
  safety, 283
  storage space, *(table)* 284
  watering system, 286
Beef cattle, 126-156 (see also Cows, beef; Calves, beef; Feeder calves)
  alternative raising, 150-153
  breeds, 130-134
  cows, 134-144 (see Cows, beef)
  dairy beef, 133
  diseases, 154
  feeders, 144-150 (see Feeder calves)
  heifers, 137, 146
    age to breed, 137
  industry, 127
  parts of steer, 144
  slaughtering, 153
  small scale, 129, 150
Breeding (see specific species; Genetics)
Brood mares (see Horses)
Brood sow (see Sows)
Butter making, 87

Calves, beef, 137-140
  castration, 138
  dehorning, 138
  feeders, 144-150 (see Feeder calves)
Calves, dairy, 72-79 (see Cows, dairy calves)
Chickens, 222-250
  anatomy and physiology, 227
  breeds, 224
  broody hens, 240, 245
  chicks, 240-246 (see Chicks)
  egg formation, 228
  layers, 230-240
    broodiness, 240
    culling, 238
    diseases, 238
    egg handling, 237
    egg production, 236
    equipment, 232
    feeding, 234

    housing, 231
    laying condition, 238
    light, 229, 232
    molting, 237
    parasites, 239
    problems, 238
    starting a flock, 225
  meat, 246-250
    feeding and housing, 246
    grades, 247
    slaughter, 248
  plucking, 249
  slaughter, 248
Chicks, 240-246
  day-olds, 240
  equipment, 241
  hatching, 245
  heat, 241
  housing, 241
  management, 243
Cockerels, raising, 246-248
Colostrum, 50
Concentrates (see Nutrition; specific species)
Cows, beef, 134-144 (see Beef cattle; Calves, beef; Feeder calves)
  breeding, 135-137
    artificial insemination, 137
    heifers, 137
    season, 135
  calving, 138
    management, 138-141
    schedule, 136
  equipment, 134
  estrus, 136
  feeding, 141-144
    guideline rations, 143
    nutrient requirements, *(table)* 142
  gestation length, 137, *(table)* 311
  heat, 136
  housing, 134
  record keeping, 141
Cows, dairy, 41-90
  artificial insemination, 65
    sire selection, 67
  breeding, 64-68
  breeds, 42-45
  calves, 72-79
    castration, 138
    dehorning, 77
    disease prevention, 79
    diseases, 78
    feeding, 72-76
    housing, 72, 74
    pneumonia, 78

    replacements, 74, 76
    scours, 78
  calving, 68-72
    interval, 52
    problems, 70
    quarters, 69
    signs of, 69
    stages, 69
  colostrum, 50
  conformation, 48
  diseases, 79-83
    bloat, 81
    mastitis, 79
    milk fever, 82
  drying off, 52
  equipment, 56, 84
  estrus, 66, *(table)* 311
  feeding, 57-64
    after calving, 63
    concentrates, 59
    dry period, 61
    grain guides, 60-62
    hay, 58
    pasture, 61
    salt, 63
    schedule, 63
  gestation length, 68, *(table)* 311
  heat detection, 65
  heat periods, 66
  heifers, 76-78
    age at calving, 77
    feed and care, 76
    housing, 54
  lactation cycle, 50
  milk handling, 86
  milking, 83-87
    let-down, 85
    procedure, 84
  milk yield, 42-45, 50-54
  parts of a cow, 48
  selecting, 45
  vs. dairy goats, 37
  weight estimation, *(table)* 58
Crossbreeding (see Genetics)

Dairy cows, 41-90 (see Cows, dairy)
Dairy goats, 91-109 (see Goats, dairy)
Digestive problems, 294
Digestive systems, 9-12
Drenching (see First Aid)
Dressing percent (see Meats)

Egg production (see Chickens)
Emergencies (see First Aid)

Ensilage (see Silage)
Estrus, (table) 311 (see also
   specific species)
Ewes (see Sheep)

Farrowing, 170-174 (see Sows)
Feeder calves, 144-150
  feeding, 147-149
    grass vs. grain, 147
    guideline rations, 148
  finish characteristics, 149
  housing, 147
  selecting, 145
  slaughter weights, 149
  weight estimation, (table) 145
Feeder pigs, 180-187 (see also
   Sows)
  equipment, 181
  feeding, 183-185
    average daily gain, (table)
     185
    feed intake, (table) 185
    garbage, 183
    grain mixes, (table) 184
    pasture, 183
    protein requirements, 183,
     (table) 184
    quantities, 183
  housing, 181
  slaughter, 185-187
    judging finish, 186
    weights, 186
  starting out, 180
  worming, 188
Feed handling, 284
Feeding (see Nutrition; specific
   species)
  formulating rations, 19, 297
  for production levels, 4
Feeding standards, 19, 297-301
Feeds (see also Nutrition)
  analysis, 12, (table) 300
  proximate analysis of common,
   (table) 300
  types, 15-18
Feed storage, (table) 284
Finishing pigs (see Feeder pigs;
   Sows)
First Aid, 290-296
  digestive disorders, 294
  drenching, 296
  general procedures, 291
  injections, 294
  kit, 290
  lameness, 293
  liquid medications, 296
  professional help, 291
  respiratory ailments, 294
  tablets, 295
  temperature, pulse,
   respiration, (table) 292
  wounds, 293

Formulating rations, 297 (see
   also Nutrition)

Genetics, 20-34
  chromosomes, 21
  crossbreeding, 32
  gene interactions, 24
  heritability, 31, (table) 32
  heterosis, 32
  inbreeding, 33
  meiosis, 23
  mitosis, 22
  populations, 29
  progeny testing, 31
  selection, 31
  sex determination, 28
  variation, 20
Gestation length, (table) 311 (see
   specific species)
Goat kids, 101-104 (see also
   Goats, dairy)
  dehorning, 103
  feeding, 101
    grain, 102
    milk, 101-103
  management, 103
  weaning, 103
Goat meat, 109
Goats, dairy, 91-109
  advantages, 91
  breeding, 99
  breeds, 93-95
  diseases, 104-107
    digestive disorders, 106
    foot and leg problems, 105
    parasites, 104
  equipment, 96, 97
  estrus, 100, (table) 311
  feeding, 98
  fencing, 98
  gestation length, 100
  heat detection, 100
  hoof trimming, 105
  housing, 96
  kidding, 100
  lactation, 100
  milk, 92
  milk handling, 108
  milking, 107
  milking stand, 108
  parts of, 92
  selection, 95
  space requirements, 96
  worming, 104
Grading meats (see Meats)

Hay, 305-308
  buying, 307
  judging quality, 306, 307
  nutritional value, 306
  plant types, 305
Hay equivalent, 17

Heifers (see Cows, dairy; Beef
   cattle)
Hens, 222-250 (see Chickens)
Heredity (see Genetics)
Heritability, 31, (table) 32
Heterosis, 32
Hogs (see Swine; Sows; Feeder
   pigs)
Hoof trimming
  goats, 105
  horses, 272
Horses, 252-279
  breeding, 276
  breeds, 256
  costs of ownership, 253
  diseases, 272
  draft, 256, 259
  estrus, 277, (table) 311
  feeding, 264-269
    complete feeds, 268
    concentrates, 267
    general, 264
    hay, 266
    nutrient requirements,
     (table) 270
    pasture, 266
    quality, 265
    quantities, (table) 269
    schedule, 265
  foaling, 278-280
    feeding after, 279
    quarters, 278
    signs of, 278
  foot and leg problems, 274
  foot care, 272
  grooming, 271
  handling, 261
  health care, 268
  heat, 277, (table) 311
  housing, 262
  lameness, 275
  parasites, 275
  parts of, 255
  pregnant mares, 278
  safety around, 261
  selection, 253
  shelter, 262
  sources, 254
  stalls, 263
  vision, 262
  worming, 271

Inbreeding (see Genetics)
Injections, 294
Iron, baby pigs, 173

Kids (see Goat kids)

Lactation (see Cows, dairy;
   Goats, dairy)
  cycle, 50
  nutritional requirements, 6

Lambs, 208-212 (see Sheep)
Lameness (see Horses; First Aid)
Light, for hens, 229, 232

Manure, 288
Meats, 112-121
  carcass weight, 114
  cutability, 114 (*table*) 117
  dressing percent, 114
  grading, 115-121
    quality, 115, (*table*) 116
    yield, 117
  marbling, 116
  retail cuts, 122, 123, 124, 125
  terminology, 114
Milking (see Cows, dairy; Goats, dairy)
Mules, 256
  advantages, 256
  selecting, 254-256

Nutrition, 4-19
  digestive systems, 9, (*table*) 13
  feed types, 15-18
    concentrates, 18
    roughages, 16
  formulating rations, 18, 297
  hay equivalent, 17
  minerals, (*table*) 10
  nutrient classes, 7-9
    protein, 7
  production levels, 4
  proximate analysis, 12
  total digestible nutrients (TDN), 15
  vitamins, (*table*) 10

Orphans
  lambs, 207, 217
  pigs, 174

Pasture, 302-305
  as feed (see specific species)
  classification, 302
  growth pattern, 303
  maintenance, 304
  mixed grazing, 304
  quality, 303
  stocking, 305
Pearson square, 298
Pigs (see Swine; Sows; Feeder pigs)
Piglets (see Sows)

Population genetics (see Genetics)
Poultry (see Chickens)
Proximate analysis, 12, (*table*) 300 (see also Nutrition)

Rations, formulating, 19, 297 (see also Nutrition)
Rodent control, 285
Roughage (see Nutrition; Pasture; Hay; specific species)

Sheep, 192-220
  advantages, 193
  breeding, 212
  breeds, 194, 196
  bottle lambs, 217
  diseases, 215
  equipment, 200
  estrus, 213, (*table*) 311
  feeder lambs, 216
  feeding, 202-209
    flushing, 212
    lactation, 208
    rations, 209
    pasture, 202
    pregnant ewes, 203
  fencing, 202
  flushing, 212
  health practices, 214
  housing, 199
  lambs, 208-212
    castration, 210
    creep feeding, 208
    tail docking, 210
    weaning, 211
  lambing, 205-208
    management, 205
    orphans, 207, 217
    pens, 200
    problems, 206, 207
  orphans, 207, 217
  parts of a ram, 195
  production systems, 194
  replacement ewes, 212
  selection, 198
  shearing, 213
  starting a flock, 195
  types, 194
  worming, 214
Silage, 308-310
  process, 308

quantities fed, (*table*) 310
  types, 308
Sows, 163-179 (see also Swine; Feeder pigs)
  breeding, 168
  diseases, 187-191
  equipment, 163, 165
  estrus, 169, (*table*) 311
  farrowing, 170-174
    equipment, 171
    management, 172
    quarters, 170
  feeding, 166-168
    rations, sample, (*table*) 166
    reproductive year, 167
  fencing, 164
  gestation length, 169
  heat detection, 169
  housing, 163
  piglets, 173-179
    castration, 177
    creep feeding, 176
    feeding, 176
    iron, 173
    losses, 175
    management practices, 173, 177, 178
    nursing, 174
    orphans, 174
    weaning, 179
  worming, 188
Swine, 157-191 (see also Sows; Feeder pigs)
  breeds, 159-163
  commercial systems, 158
  diseases, 187-191
  feeder pigs, 180-187 (see Feeder pigs)
  parts of a pig, 191
  piglets, 173-179 (see Sows)
  sows, 163-179 (see Sows)
  space requirements, 163, 181

TDN, 15

Veterinarians (see First Aid)

Watering systems, 286
Water quantities consumed, (*table*) 287
Worming (see specific species)
Wounds (see First Aid)